THE TECHNOLOGY OF BELIEF

James True

Copyright © 2020 James True

All rights reserved.

ISBN: 9781697181289

*Dedicated to
John Wayne Anderson*

I see him pushing his glasses up his nose now as he's about to speak. He would aim his head sideways like a gangster when he listened. The truth never got away from him. Most days of his life, John was pulling squashed souls out of the grill of the machine. He was dirty, like a good mechanic. He held a confident posture on pine crutches. He never showcased the pain. Suffering was an ingredient for his medicine.

1 - The Technology of Belief ... 1
2 - Moon's Field Notes ... 3
3 - Oracle at Delphi ... 7
4 - Needles of Cleopatra ... 13
5 - Medusa of Gorgon ... 25
6 - The Electric Cobra ... 30
7 - Behold a Pale Horse Ass ... 36
8 - Blackmail and Whitemail ... 43
9 - Alchemy of Airships ... 51
10 - A Smooth Criminal ... 55
11 - Fire & Isis ... 61
12 - The Satanic Messiah ... 67
13 - The Trojan Horse of Zionism ... 75
14 - The SDK of Magic ... 84
15 - The Man from Katuah ... 91
16 - Equality is a Bad Word ... 96
17 - The Snake Oil Messiah ... 100
18 - America Believes ... 107
19 - Secretions of the Spider ... 109
20 - Corporate Pride Month ... 118
21 - Sins of the Father ... 123
22 - Flat Earth Karate ... 127
23 - Trumps Flow State ... 132
24 - Billion Dollar Liars ... 138
25 - Definition of Evil ... 143
26 - CNN is the Government ... 146
27 - The Prana Economy ... 153

28 - Government is Mafia ... 161
29 - The Wasp and the Caterpillar ... 171
30 - The Second Coming ... 181
31 - The Capital of Punishment ... 189
32 - The Two Towers ... 195
33 - Apocalypse Now ... 221

CHAPTER ONE
The Technology of Belief

There is a technology to belief. Ancient ideas have been unplugged and hoarded. We toil to complete the circuit. When a circle is fulfilled, the ground glows. Our shoulders buzz like filaments when someone shares truth. We are swimming in plasma. Our lungs are gills in an ocean.

Belief is the aether endowed by a flock. Our beliefs have been enslaved for centuries. This happens in religion, science, and politics. The power of belief is always mistaken for its costume. We only give credit to its props and choreography.

It's a statistical fact that half of all scientific research will be proven wrong within twenty years. Still, we believe in science. It was shown recently that two-thirds of clinical studies couldn't be duplicated. Still, we give science every benefit of our doubt. We dismiss belief as childish. We coddle science like a pimp. We pretend all legitimacy is found on the surface. But below language there is sound. Below sound, there is intent. Below that, there is a belief technology.

Kabbalah and Enochian magic both tap into belief. These are ancient arts charged from the power of reverence. Tradition is a tangible force in belief. The rarity, age, and complexity of a system yield tremendous torque. One who masters its intricacies draws prana from the energy of awe. Belief is a measurement of electrical confidence. A mother knows she can lift a Buick to save her child, so she does. She speaks no incantation. She follows no ritual. She instantly converts her body into cortisol in a holocaust of sugar. Her sacrifice in the face of her child's death is an economic exchange. She signs an instant contract with the aether.

This book is an appraisal of the technology of belief. There is a powerful science hidden in our life-force. These chapters demonstrate its effects through history and the future.

CHAPTER TWO
Moon's Field Notes

Words are poor as a poet. All of our thoughts are slippery fish flopping on the dock. We put our hooks in their gums and pull them out to show each other what we thought. Meaning is a colored koi squiggling through our fingers. We lose its intention pretending to hold oceans in cups. Utterance is a drop in the sea of every thought that ever was. We are neurons on the omnidirectional train tracks of prejudice. God is zero-point back at the station. Our identity is a vector as much as it's a pulse. We are surging streams through time and space. If you want to know how God sees you, take a snapshot of every pose your body's ever held and connect the dots. Make a map of every geo-coordinate you've occupied. What does your time-body resemble in it's true form? We are time squiggles. Like three-dimensional ink from a calligrapher's pen, our spline tapers, splotches, and zips along a baseline. We are a song of the deep for the highest of places.

The meaning gets lost in translation. We miss each other interpreting our giftwrap. We are squirrels arguing with a tree over the shape of the breeze. Language is confusion masquerading as comprehension. We reward ourselves by telling each other what they should have thought.

When the love of self goes silent, there is apathy. When apathy goes silent, there is love. We are emotional lungs filling and spilling from the tide. We cast God out of the garden and accuse him of banishing us. These charades have perfected the art of victimhood like a firecracker. The apple of truth is a psychedelic flesh. Eve's fruit was the ultimate red-pill. Before that first bite, there was no veil. Our lips are still puckering from the tang of morality's dimension. Our tongues were virgin before the first lie. We held our insides and outsides in a similar esteem.

At sea, the difference between flotsam and jetsam is intent. In maritime law, what we lose at sea is different from what we abandon. We live in a dimension of intent. If flowers were human, they'd find a way to resent the bumblebee. Appreciation is the long stamen in a womb of petals. Morality is invisible to flowers. Morality is real as war to humans. We are seven billion hands clutching a single gold compass.

The moon will never turn from its crop. It's up there waiting for you to hatch. We crack open when we die, and the moon catches our field notes. Every picture we make is placed on the galactic refrigerator. We are solutions rendered in cranial fluid. We secrete sweaty potions that kill us slowly through time. There is so much to reveal in the underworld. But that part is a grand surprise. Anticipation is a thorn under the silky skin of a rose.

You know you're pure when vampires love your juice. They want your body, and you feel alive when hunted. Rabbits beg for the chase by twitching their nose in the bush. Let us speak true of the deeper shade of soul. Underneath the yearning for survival lies a passion for death. How much of yourself do you know? You are naked after a shower lying on your towel. Your hair is wet, and your teeth are dry as you contemplate the swing of a ceiling fan. The shadow grows long before it drowns. The orchid drops its spent sculpture at night when no one's looking.

Are you ready for the underworld? Crawl under your house and see. Feel the cool slippery plastic as you scooch your belly like a snake. Reptiles breach the ground with their undercarriage. Their keels crest the inverted waves of Hades. You will never know thyself till you see your shadow.

CHAPTER THREE
Oracle at Delphi

In 1400 B.C., a young man named Kamesh, led a trip of a dozen goats along a craggy ridge-line of thick laurel overlooking the Mediterranean. Under a blue sky, they marched on the green fingers of rock spilling their pebbles into a cobalt sea. The crescent bay they just found would soon be sacred ground. Geology split the earth in a jagged scar making access difficult. Kamesh found a flat knoll where his bearded tribe collected fresh grass in their teeth. One of the goats ventured beyond the knoll and Kamesh went to fetch her. In six jumps, the goat claimed a tall chimney of rocks too steep to reach. Leaking cracks of ethylene gas poured from the ground making Kamesh's thoughts blurry. He could not coax his beast home. Kamesh, now enthralled by the gas, threw rocks to drive it down. Intoxicated in the ground's sweet aroma, the goat bleated a song of defiance and revelation. Kamesh watched it call forth the Copper God from the Golden Dawn. The goat was summoning Apollo. He would claim these rocks as the temple at Pythos, the site later known as Delphi.

For the next millennium, people pilgrimaged here to commune with the Oracle. The goat was replaced by a young sweet virgin groomed from the local village. She was named the Pythia, voice of the Pythos, the stern lips of Apollo. She became a living archetype of intuition. The Pythia answered questions from pilgrims at the waxing of every moon. Above the entrance of the Oracle's temple, it read in Greek, "Know thyself." The legend grew as people would come from hundreds of miles for a chance to consult the Oracle.

The Pythia was made famous by Homer and conferred by many Roman emperors. To seek the Oracle was to surrender one's reason to the power of Apollo's intuition.

Seven sisters drew Caspia's hair from a copper basin of milk and honey. The brigade of giggling maids spent the evening weaving flowers, seashells, and berries into ceremonial tinsel. They spun her mane around a crown of laurel. They wrapped her virgin body with soft linen and draped vines of garland behind her back and through her elbows like a shawl. Caspia was the new Pythia. She was chosen at sunrise in the taking of auspices. Two Cretan priests saw starlings split into three, then again into two. Caspia's destiny bloomed as the second child born from the third daughter. She was the virgin bride chosen by Apollo at Mount Parnassus.

Caspia officially became Pythia before dawn in a ritual bath. She rose anew with the sun into the seat of laurel atop a tall copper tripod. She untwined her spine in the temple's slow pulse of adoration. The ritual believed in her so she did, too. Belief played her endocrine system like a deep drum. Rapid breaths pushed her deeper into the trance. The ethylene was a kind of brain death. Creams of sweat gathered to drip from her chin. Her brain's theta entrained with its gamma waves. For the next day, Caspia was a passenger of her intuition. Her lips answered questions before the sound met the perimeter of her ears. She spoke wisdom on things she could not have understood from positions she could not relate. One morning, she stopped a King mid-sentence with the confidence of a leopard. She had burned all doubt at the stake. In the market of tongues, we see our power in the silence we command. To hang on someone's word is a sign of obedience. Attention is loyalty. Caspia was the living Oracle. Her rational mind was too frail for such a vessel. The Temple of Apollo consecrated her gut from a flight of birds and a goat's song. The power of belief flows like electricity through a girl on a copper throne. The Oracle at Delphi survived for a thousand years. It was vital belief technology harnessed by pilgrims all over the Mediterranean.

Like any church, the Oracle was corruptible. The technology of belief creates a vortex for feeding and predators form like flies. Psychopaths gotta psychopath. Corruption is the wilting of civilization. The young Pythia was replaced with a secret committee. Pythias were elevated by their ability to obey and perform. The golden child was marketed to the masses. Corruption is a long way down.

The modern-day oracle qualify himself on paper. They dress for a ceremony in a white collar and a silk tie. The modern suit bears the trappings of the Cretan priest. Our oracles of justice wear black robes. Our oracles of medicine wear white. We are entrained in the essence of costume.

We consult what we believe. If the Oracle said one would prosper, the pilgrim left with a prosperous belief. They were supercharged by the virgin on a tripod. The effects of the belief would be measurable. Confidence walks in a light of divine assurance. It meets the pilgrim every morning as they step out of bed. The professional athlete wears the same socks to win a game. An infantryman survives a war with a girl's picture. We seek our oracles subconsciously every day of our lives. We look for something outside ourselves with enough power to overrule the intellect. We are the face behind the mask our power can't afford to see.

Fooling ourselves is the purpose of divination. From the flight of birds, tea leaves in a cup, cards in a spread, or flips of a coin – chance is a technology of surrender. We tell our intellect to sit down as we abandon the will to fate temporarily. We hand ourselves a verdict empowered by an investment in something outside ourselves. This yields a profit of confidence. This is why we consult experts in costumes. We are jacking into the belief. This is why we only seek counsel when we have doubt.

Oracles are a well of psychic empowerment. At the core of all magic is confidence. In 350BC, Phillip II of Macedon, father of Alexander the Great, went to see the Oracle. He asked if he would defeat the Persian King in the east. The Pythia waited several minutes for her intuition to breach. She looked down on the king and said, "The Bull is crowned. All is done. The sacrificer is ready." Phillip left with fate's promise of victory and he won against the Persian Army. But the Oracle's decree came in spring when Phillip was stabbed by his ex-lover.

The Oracle at Delphi was influential only because of belief. No army enforced her decree. Delphi was a temple for intuition. We still tap the same technology masked in consumerism. Consumerism is magic. Corporations package beliefs and sell them back to us. It works because we unlock our worth through a purchase of energy. We buy prana now. It's shrink-wrapped in fantastic plastic. We only harness a small percentage of our own power after it's drained by the system. We believe in consumerism and the power of products. We market them to each other as "systems."

This is why packaging is so important in this world. The facade is the form and the function. We believe in Alka Seltzer because the box is shiny. We believe in Pepto Bismal because it's pink. The power of the hive generates a placebo effect. From new cars to athletic shoes, products are the capital of belief.

Instead of gassing a perched virgin for advice, anoint yourself Pythia. Direct the energy of your confidence back into your body's temple where it can be seated. Man is a trinity of compassion, reason, and intuition. You were brought here to tap into its expertise. The more we surrender our power, the more we live as slaves. Tell the moon it's your last ride. Make your life an alchemy to inner-confidence.

CHAPTER FOUR
Needles of Cleopatra

There is a science to the obelisk. Originally called tekhanu in Egypt, the four-sided, monolithic symmetrical towers always taper at the top to form a pyramidion. Pyramids and obelisks bear measurable effects on atmospheric electricity. For every meter above ground, the conductive antenna generates 100 Volts. Energy flows from the earth up the obelisk and radiates into the atmosphere. This movement is a flow of electricity inside a toroidal field. Each obelisk is an energy torus drawing from its core and spilling out into the air like a fountain.

Egyptian obelisks are cut from piezoelectric granite. Piezoelectric materials are organic batteries you can charge with pressure or heat. Natural crystals like quartz, granite, topaz, and even table sugar have piezoelectric properties. So too do biological materials like bone, silk, wood, and DNA. Piezoelectric quartz crystals powers every electronic device we see today. The cleavage of the crystal determines the direction and reliability of its pulse. The notion that crystal power is psycho-somatic comes from people with ignorance of electronics. People who mock the power of crystals are idiots.

Obelisks are piezoelectric devices that absorb vibration and eject energy into the atmosphere. This science is documented. Suspending a wire forty feet in the air can power a low-torque motor. If the top of the wire terminated into a sharp point, like a pyramidion, you would create even more energy. The symmetrical point, or capstone, found at the top of obelisks, pyramids, and domed roofs focus the electricity. This is the science of grounding.

The obelisk is an archetypal landmark in the psyche. Like the world's tallest building, obelisks come with an energy of reverence. The power of experience is easily lost in the age of video and CGI. We lose the tangible electricity of experience versus seeing its picture. Experience and pictures are two very different things.

The ritual of charging the ground below an obelisk with blood was popular as early as Delphi. For a thousand years, a monolithic stone called Omphalos was ritually anointed with oil and wrapped in wool. The stone was cherished as the navel of the world marked by two eagles measuring the sky. Omphalos was a living stone planted in sacrificial blood. The life charged the stone and made it a vibrational marker. Today, we still plant monoliths above a carcass. Dead stones are planted from the dead. Living stones are planted from the living. Sacrifice is the conversion of energy. Like thirty-eight-trillion fat cells burning simultaneously. The monolith is charged with a single note. That note is the intent devoted to a single purpose.

The power of living stones is a mostly forgotten technology. Rome celebrated the god of boundaries, Terminus, every Feb 23rd. It was a ritual death of the year. Landowners participated in the consecration of property markers with a sacrifice of living blood. Living stones were considered binding contracts sealed with ritual witnesses.

Three-and-a-half thousand years ago, in the city of Heliopolis, Pharaoh Thutmose III was glowing in the starlight. Children tasked with aiming copper shields dramatically lit their Pharaoh from below. They scurried and chased Thutmose like ballboys at a tennis match. Each concave shield held a small oil lamp in the center. The light was collected in the parabola and reflected onto Pharaoh from every side. This kind of attention was all Thutmose had ever known. The Pharaoh was a living 24-7 reality show. He was the apex for human belief. A deep confidence surged through his veins that deeply nourished his surroundings. The dynasty of Thutmose knew the secret of mind control. They convinced a population Pharaoh raises the sun with the palm of his hand.

Two builders approached Pharaoh presenting his golden merkhets and sighting rods. On Pharaoh's order, they found the constellation Callisto and traced her saddle across the back of the dome to Polaris. A priest approached Pharaoh with the golden cord and waited for his command. Pharaoh marked the final orientation of both seventy-foot-tall obelisks. The pair would align to his will and create a portal to the temple of Thoth. In the ground, below each foundation, a pit was dug and filled with offerings of tools, jewelry, and statues of the ibis and baboon. With his right hand, Pharaoh poured sand from the western shores of the Nile into the first pit. With his left hand, he poured sand from the eastern side into the second. The stone frame of the portal was consecrated by the shores of Egypt's sacred river. As below to above, a doorway formed from the shores of the Nile to the beaches of the Milky Way.

Two bronze chairs were positioned above the pits. It was time to add the blood offering. The ranking priest rose to the pulpit and began a call for Holy Oil. The congregation repeated the chants of the priest like the chorus at a concert, "Sheshmu come through." Sheshmu was the god of oil embodied in the body of a lion. He was bigger than a war elephant and revealed himself in the hot mist of freshly evaporating sacrificial blood. During the ritual, the blood is adrenalized and the brain ceases to inhibit DMT. The human sacrifice enters a trance of epiphany and the Holy Oil can be extracted. Before sunrise, his life-force spurts into the foundation.

Epiphany is induced through a secretion from the pineal. This gland is buried deep in the heart of the skull. It infuses the circulatory system with metatonin, a hyper-potent version of melatonin. It's an electrical orgasm for consciousness. Like yeast rising in the oven, you can't pause or store epiphany. It's a state of transmutation. Under epiphany, the psyche can possess any archetype he desires if he knows how. Sheshmu was the embodiment of raw psychic will and power. The sacrifice would channel Sheshmu to give Pharaoh all of its prana for the charging of each stone.

To find a volunteer, the priest would squeeze the congregation into a climax until someone surrendered. If no one stood up, the priest would be offering his own life in their place. This made the ceremony elaborate and manipulative. The priest employed an arsenal of musicians and provocateurs embedded in the crowd. They orchestrated giant paw prints in the sand and dramatically witness the hot wet heat of a cat's breath passing overhead. In time, the drums would surge as the priest cried for the rising of Sheshmu. In a crescendo of pounding he called, "Who gives oil for Sheshmu?" At the moment of climax all the fires were doused with oil. As members of the congregation jumped from the jolt some found themselves selected by their neighbors. At times, a man would stand from his own volition. Often, a pariah was pushed into the aisle by a coalition. The chosen were raptured quickly upon shoulders and ushered to the front to fulfill the promise. The priest would only pick two. It was an honor to be spared if you begged to be taken.

Two sacrifices was drawn and quartered by attendants to lend them courage. The priest gorged his knuckles deep into their groin to block the pulse of the femoral. A sharp slice from a blade opened the thigh as the priest pulled out his artery like a spigot. The patient was seated upright in the throne for draining. He is raised above the cheering crowd as his Holy Oil pulses and spurts its way down his leg to anoint the foundation. The victim is seated in a throne of hot blood as his shoulders shiver in the clammy shock. His vision dilates to reveal the golden mane of Sheshmu rising over the horizon to greet him. His dying thoughts are a blessing to Ra's needle. For eternity, his blood will embody the obelisk's foundation.

After the oil is drawn. The needle's erection begins with the rise of Pharaoh's hand. He called the sun. The tall twin needles rose from his utterance. The belief of seven hundred men and twenty-six elephants lifted the pillars in line with the Aten. By Ra's zenith, the living twins shone like lightsabers. Each pillar was coated in brilliant electrum. Thutmose consecrated the doorway's threshold as the first to enter. Between the pillars, he decreed the temple to Thoth would stand forever. But forever comes quickly for Cleopatra.

Fourteen-hundred years later, Thutmose's promise was broken as she sent his needles to Alexandria. The electrum was scavenged long ago when Heliopolis turned into a boneyard. Cleopatra was Greek, not Egyptian. The courageous Sheshmu was long forgotten and the city of Alexandria wanted the needles to seduce Rome.

When Caesar defeated Pompey, he restored Cleopatra to power. The couple took a cruise up the Nile to the Temple at Luxor. Along the journey, Caesar relished the way Egyptians treated Cleopatra. Rome respected Caesar but Egypt was enamored by their Queen. She held a divine charisma fueled by reverence. She was a graceful goddess who parted the waters when she walked. Caesar ruled people from their mind. Bribery, fear, and promises are the tools of politicians. But Pharaohs ruled from deep devotion. Cleopatra would take Caesar to Karnak to show him what he was missing.

Devotion can be measured through self-mutilation. The practice of circumcision arose from Khemit. The ritual was adopted by the earliest kings of Egypt. A young Pharoah was rewarded with dopamine for the show of strength while his genitals were sliced with a dagger. So say the shepherd, so says the flock.

In mutilation, the flock is rewarded for body disassociation. The subject is controllable when the body is evacuated. This is the essence of possession. Egyptian alchemy is the manipulation of dopamine and serotonin. A public reward for self-mutilation breaks a man like a horse.

At Luxor, Cleopatra introduced Caesar to the secret teachings. Inside the Temple of Man, he was initiated in the secret society of Pythagoras and Aristotle. Caesar was taught the golden section, the magic squares, and the sound of divine cords. He was taught the structures of the brain, the heart, the gut, and the workings of the pituitary and the pineal. The rising of the Nile was an analogy for the body's natural chrism. Caesar immersed himself in the magic of Osiris and learned to rule from the crook and flail. He spoke new spells of milk and honey as he studied the science of pheromones from his Queen. For months he dove into Egyptian sex magic and gave Cleopatra his seed. The couple left Karnak on the closing of the feast of Opet; the mother of Osiris. Cleopatra had timed her sex ritual with the triad of Amun and Mut. The trinity of Caesar and a pregnant Cleopatra emerged from Luxor between two granite obelisks. Back on the Nile, she told Julius, "You came to Luxor with power. You leave with divinity."

Cleopatra destroyed Egypt's energy by moving the needles. She broke the portal's seal and plundered its reverence. The ancient kingdom was cannibalizing itself. Rome has more of Egypt's obelisks than Egypt. Only six remain standing in their original position. The twins from Heliopolis along with the east obelisk at Luxor came to be known in modern times as Cleopatra's needles. In

1833, King Louis-Philippe placed the Luxor Obelisk on the same ground as Marie Antoinette's guillotine. The original blood ritual reverberates.

Forty years later, the remaining twin needles set sail to New York and London. Six people drowned at sea as one sailed past Gibraltar. It was later erected in London on the base of a sun scarab. Three years later, the New York needle rose in Central Park by the Vanderbilts. It sat atop a base of crabs. Carter Vanderbilt would jump to his death as the sun rose in Cancer. He died a few blocks from Cleopatra's needle. It was Vanderbilt's needle now.

French celebrity chef Eric Ripert posted a photo in front of the Vanderbilt needle four days after finding the dead body of his celebrity friend, Anthony Bourdain. Bourdain's death was recorded a suicide but the circumstances fit a human sacrifice or touchless kill. Humans are living obelisks and the golden child is always plated in electrum. Ripert's photo of the needle was captioned, "Mysterious ,fascinating and beautiful obelisk; only revealing its hidden secrets to the initiated and knowledgeable ones."

History charges relics as much as magicians do. The crystal matrix of the obelisk is a battery for prana. The technology of belief interfaces with its power. Attention is an energy with voltage. The obelisk in St Peter's Square is one of thirteen taken from Egypt. In 37 BC, Caligula brought it from Heliopolis and stood it erect during the martyrdom of Peter. The saber decorated the persecution of Christians and witnessed the burning of Rome. Pope Sixtus officially exorcized the stone before moving it to the square and crowning it with a cross. Nothing erected in the Holy City eighty-three feet tall and weighing a million pounds is an accident.

In 1586, nine hundred men and seventy-five horses raised the petrified ray of Ra. Sixtus inscribed the obelisk, "Christ conquers, He reigns, He commands; May he defend his people from all evil." On the apex of the needle, a bronze cross rises from a morning star and tunes the energy of the obelisk. The stone is a quartz drum that rings like a bell. Shards from the True Cross are sealed inside the bronze acting as a lighthouse beacon. The obelisk becomes a living altar of esteem as we venerate a human sacrifice. Rome runs on the secretions of crucifixion. Rome conquered paganism with human sacrifice. Rome gelded the intuition with shame and the belief in damnation. Every Egyptian obelisk in Rome has been conquered by the cross.

Constantine saw with his own eyes in the heavens a trophy of the cross arising from the light of the sun, carrying the message, In Hoc Signo Vinces or 'with this sign, you will conquer.' - Eusebius.

Religion is the bureaucracy of belief. The Holy See is a hungry leviathan and Christ is the passover lamb. The obelisk in St Peter's Square is a Terminus contract erected shortly after the church claimed, "the necessity of salvation for all Christ's faithful to be subject to the Roman pontiff." The Vatican needle stands in the center of a zodiac. The pagan traditions were conquered not incorporated. The cross was planted in the ground like a flag on the beach of some foreign land.

The archetype of the Apostles is possessed by the houses of Black Nobility. There are ancient families standing in the Saint's limelight. They are empowered and venerated by a story of martyrdom. Every Saint's death is more gruesome than the last. Black magic rides the coattails of sympathy and reverence. Human admiration is an engine run on esteem. The gates of St. Peter are made of hoarding shoulders crowding the light. The Oracle of Rome is an Egyptian phallic stone delivered by Caligula. It's apex was circumcised by a bronze cross atop a morning star. The church trades our prana for the deliverance from its evil. It is the addiction and the cure in the very same venom.

CHAPTER FIVE
Medusa of Gorgon

People think Theism is a belief in God. People think Atheism is a rejection of God. But it's closer to say theism is a particular kind of God and atheism is a rejection of that particular kind of God. The God of theism is a God who both created the world and exists outside of it. At first, this seems harmless, but there are teeth behind those lips. A God who exists outside of the world leads some to think the aether is godless. This idea is venomous drool dripping from the fangs of materialism.

Medusa of Gorgon was a child from Heliopolis. Her family roots granted her a position as Virgin in the Temple of Athena. Her favorite duty was tending the parliament. A dozen owls were housed and fed inside the Parthenon. A roost was made in the right hand of the statue to the goddess. Medusa climbed a rope every morning made of horse mane and river stones to reach them. She was nimble up high where the other girls were nervous. No one wanted the honor of tending the owls. The job was gross and required live feedings. Medusa would gut mice gladly with a seashell and peel them open like shrimp for the younglings. The fresh organs of rodents were an offering to Athena. Medusa felt the owlets were the embodiment of the goddess. They spoke to her sometimes. The painted plaster statue never would. She didn't like how many people came to worship a painted statue.

The towering Athena featured a winged Nike reaching out from Athena's hand to crown the goddess in victory. Medusa didn't like Nike either. Victory was the antithesis of wisdom. Athena is the coming and going of doubt and inspiration. Victory is an acquisition and the tide can't be carried home in a golden cup. The true spirit of wisdom moves in the silent wings of the owl who's always hungry for gnosis. This philosophy made Medusa a lonely girl in her coven. As her isolation grew, Medusa became one of the owls. They winked when they shat on the goddess. She rather enjoyed cleaning up after them. She felt like she was rubbing it in. The blasphemy of her lonely giggles gave her goosebumps. They sent chills down the backs of her sisters. Still, no one in that temple loved Athena more than Medusa.

On a misty afternoon in the house of Scorpio, sad men came from the sea carrying the fresh corpse of their governor. His final request was an Athenian ceremony of cremation. Medusa and her sisters would serve as ritual mourners. After the burning, the governor's bones were collected and placed in a burial urn. Medusa poured sacred oil she had pulled and pressed with her fingers for hours. Months of work from the olive trees were sacrificed instantly over his remains. She felt the loss and knew this made the magic work. She poured herself into his passing. By sunset, the burned dust of a governor was deep in the dirt under a large flat stone. That evening, the ceremony of the phallus was performed. Medusa ritually stroked and rubbed oil on a pillar adorned with a scarlet bow. This is a sensual ritual designed to arose the gods and entice them to bless the newly dead. Medusa called each god by name into the pillar. She anointed the wooden obelisk with passionate strokes of oil. She begged for the gods' blessings while she tended them. The ritual was designed to harness and captivate both the gods and the participants. Attention and arousal are a Holy Oil and Medusa was a worthy vessel.

The following day, a feast began on the day of Saturn. That night, drunk men saw the ripe Medusa descend from her ladder after checking the owls. One grabbed her neck while others took her wrist and ankles. They pressed the struggling Medusa into position and stuck their forks in her like cooked meat. Her face was pressed into the plaster of the goddess as men dug through her clothes. Medusa's was spread and plundered as snakes from the sea spit their salty venom into her parthenia. Medusa was abandoned at the foot of her goddess once the men finished. She abandoned herself in the trauma. The real Medusa was never seen again. Inside her skin, she ran from her pelvis. She climbed as high as she could to slither out of her skull like slithering snakes of keratin.

The temple mother would find Medusa coiled in a pile of shed skin. She ascertained that Perseus, the son of the dead governor, was her attacker. His bloodline meant Medusa would be expelled from the Temple of Athena. The Gorgon was outranked by the Perseus. His golden shoes came with the perks of the golden child. There would be no justice for Medusa and Athena was the witness.

A spirit of vengeance picked up Medusa as she rose from her pit like a viper. She met Perseus on his boat with a dagger but he was quick and caught her by the mane. He made fast work of her intestines with the very blade she brought him. Medusa's guts spilled to the deck like eels from a barrel of wine. Perseus ordered her head severed and mounted to the ship's bow. Her guts were nailed atop her skull as the rest of her body was abandoned to the waters of Poseidon who said nothing.

Most of the brain is wired for facial recognition. One person's pain is broadcast into another person's nervous system instantly. Emotional communication is empathy at the speed of light. We smile from each other's smiles. We yawn from their yawns. We cry from other's tears. Staring into the frozen face of Medusa emits a pain in high definition. The body feels what she felt and it turns us to stone. We can only break the spell of Medusa with a reflection of compassion. Evil is a cold machine selling men victory instead of wisdom. Materialism is a worldwide crusade turning everything to stone. We petrify the living aether one mind at a time. Turning someone to stone is easy. Getting them to come back is a different kind of magic. So is the truth of Medusa.

Reject the vampiric psychology of materialism. You do not live in an ever-expanding vacuum of spin. A place where no sound can ever be made or heard. No up or down. No mark of center. Where God is pushed outside a void of exploding refuse. Reject the myth of cold black space that boils the blood in an instant. Where you run to find God in a cavern of dark energy that's deeper than the speed of light. Awaken from that nightmare. Life is the breath of God. Matter is spoken word. Love is endless song.

CHAPTER SIX
The Electric Cobra

Petroleum is the most energy-dense practical fuel society has ever known. It wins in every category from renewability, portability, and storability. We have an emotional aversion to petroleum. This feeling was injected by programmers. Petroleum is not crude oil. Petroleum is abiogenic. It's not a byproduct of any living organism. We have been manipulated emotionally. Our thoughts are herded by propaganda that nips at our heels like sheepdogs. We are charmed by the electric cobra.

The electric cobra is so big we don't see its true form. Its a lie behind the enigma. Media and government are Dr Jekyll and Mr Hyde. The Mueller investigate ran for two years requiring forty agents, 2,800 subpoenas, five-hundred search warrants and five hundred witness depositions. Over 500,000 news stories were pushed to the public by the AP, Reuters, MSNBC, FOX, CNN, NBC, ABC, CBS, and the NYTimes. All of these stories were offshoots of the original Love Trumps Hate campaign. None of the stories were based on factual evidence. Several reporters were awarded the Pulitzer. Politicians and news outlets are emotional terror cells. They bring rattlesnakes into camp twisted around sticks and brandished like torches.

"We only have 12 years to live." – *Alexandria Ocasio-Cortez*

A boa constrictor suffocates its prey. The sociopath keeps it alive to feed on its suffering. It becomes an oxygen dealer. It turns something the victim needs into an addiction. We are slaves to the masters of information. We are spoon-fed for the competition of who swallowed it the fastest. We've come to expect our news refined like strained peas. The soft cool lump of glug on the tongue barely needs a swallow. We stare blankly into our mother's eyes while it oozes down our throat. We are entranced by the convenience and melodrama. We find comfort in the mythology of scientism.

"The world is no longer at risk of running out of resources." – BP

We find oil six miles below the ground. This is miles below the oldest fossil record. Oil fields all over the world replenish themselves naturally. Hydrocarbons are synthesized abiotically. Petroleum is a natural lubricant chemically formulated by the earth. People who believe oil comes from dead dinosaurs are misinformed. There are a lot of misinformed people out there. They were made that way on purpose.

If people knew oil was renewable there would be no energy crisis. Without crisis, you have no solution. Without chaos there is no order to champion. We insist oil is crude. We say over-and-over it's not renewable. We have underground strategic oil reserves and prevent domestic exploration and access. Alaska is home to the largest oil discovery in the past thirty years but its controlled by a foreign conglomerate. The earth gives, the controllers take.

We invented fracking and Flint Michigan to push a crisis and introduce a solution. The solution is always the electric cobra. The pyramid in plain sight. We are misled by emotional terrorists posing as government and the media. Scarcity breeds profit and desperation breeds decision. We have to stop regurgitating things as if they were true.

Like babies hypnotized by a face, the game of peek-a-boo is a scarcity program. The vagus nerve wrapping the heart, lungs, and gut constricts with stress from our disappearance. The adrenals embrace joy and relief when the face returns. Like Pavlov's dog, we crave the anticipation. We extend our pain to receive it. We are chemically programmable to the drama of relief.

The truth is we bless the ground. We are a walking fractal of earth. We are as important to the ground as blue whales are to the ocean. Mind control calls us a parasitic virus. It keeps our life-force subdued. Prana is an ego beating like a heart. It's a churning dynamo of bloodline. You stand as the genetic expression of your bones. You are the living spokesman of your ancestors.

When meteorologist and founder of the Weather Channel, John Coleman called climate change the "greatest scam in history" he was rejected as crazy. When the co-founder of Greenpeace, Patrick Moore speaks out against anthropomorphic climate change we reject him as a self-hating zealot. For decades they were heroes cast out of the tribe for doing what they always did – share their passion about climate. They are heretics now. Ad hominem is a prime symptom of mind control. We live in a nation of electric voodoo.

"Amy and I are the last great hope for America." – Beta O'Rourke

Al Gore's movie, *An Inconvenient Truth,* tried to convince us we are parasites destroying the earth. None of the science was true. His movie won two Oscars. Believing these people doesn't make you environmental – it makes you uninformed. Motivated people who are uniformed don't care about the environment. They care about vanity. Fame is a prime mover in the land of the spotlight. Greta Thunberg is the new season's David Hogg. Today's climate protest is a public relations decision with corporate sponsors. Demanding action is the same as demanding government to grow. We push ourselves over a cliff as we warn each other about the fall. The crowd is simply too ignorant to notice. It's why government grows fat when it's stuffed with people. The larger the organization, the more it's okay to lie.

Theft becomes a norm in organizations with more than eight people. The corporate system causes morality to recede. This same corporate system dictates our war policy and vaccine schedule. We are surrounded in corporate psychopathy but since our morals are muted we think it's okay. This is why climate scientists fudged data with Al Gore. The same reason NASA was caught reporting false sea level rise in 2017. The larger the pyramid the bigger the umbrella for evil.

When we delegate authority we dismiss impeccability. Morality is extracted from our autonomy. Immorality is a byproduct of slavery. The modern man can only afford to be sovereign on weekends and two weeks in the Summer. The rest of the time people are just doing their job.

Look at our legal system. In 2011, we had over a million lawyers in practice. Every one of them was impotent to end a company like Monsanto. 9/11 billionaire profiteer Larry Silverstein is not under litigation from 2,996 separate civil lawsuits. Not one member of our government has faced investigation or trial for negligence. On that September morning, everyone did exactly what they were told.

"It is a strange game. The only winning move is not to play." - War Games

Ask a lawyer if they defend people who are moral. The honest ones will tell you no. Lawyers are hypnotized by a belief in the system and litigation leaves them fearful of the reprisals. Not one of our judges represent the people. Their humanity has been cloaked in a black robe. The legal system is a court of dissonance. Justice is a corporation fed by apathy and silver.

CHAPTER SEVEN
Behold a Pale Horse Ass

Alex Jones is no Bill Cooper. Through Bill Cooper's writings and broadcasts, we were told how the system worked. Bill told us they would come for him one day and they did. Bill Cooper was wanted by the Clintons and the IRS at the same time. He was named a major fugitive by the United States Marshals Service. Bill Cooper died two months after 9/11 in Eagar, Arizona. Two deputies came for him and Bill Cooper died in a gunfight. It could have been a lazy afternoon. But Bill hadn't paid his taxes. Bill was smeared by the law, the courts, the IRS, and the Southern Poverty Law Center. He predicted and conveyed the truth. He warned us of what was to come. He brought home the blueprints to the Death Star and the American government killed him for it. Our taxes shot Bill Cooper. Our silver turned the gears of a federal meat-chipper that hunted him down and ate him. His book, *Behold a Pale Horse,* is scripture for anyone wanting to understand the capabilities and motivations of the federal government. Bill taught me countries aren't what we think they are. Bill taught the key to control is the illusion of dominance. For any authority, nothing is more important than controlling an illusion. The quest for truth becomes an endless round of professional wrestling. The house wins when everyone buys a ticket.

Evaluate crumbs: Usama Bin Laden gave CNN a live interview in 1997, ABC in 1998, and Time in 1999. Since 1990, UBL was a high-value target of United States intelligence and Mossad. Both agencies are embedded in the American media and both missed three opportunities to apprehend him.

On June 28th, 2001, Bill Cooper predicted something big was about to happen. He predicted they would blame whatever it was on Usama Bin Laden. Ten weeks later, on 9/11 something big happened. Two months afterward, Bill Cooper was dead. The pale horse Bill Cooper wrote about was truth in the clear light of day.

Bill Cooper is a Pale Horse. Alex Jones is a Pale Horse's ass. Alex Jones interviewed Bill Cooper on his show in 1998. Bill Cooper called Alex Jones a fraud. Bill knew the dangers of someone like Alex. On New Year's Eve, 1999, Alex Jones claimed Russia had launched inter-continental ballistic missiles at America. He covered the false event for hours. Alex was demonstrating his addiction to fame. It overrode his reputation or quest for truth. Addicts prioritize addiction. This makes them predictable. Anyone controlling an addict's supply controls their mind. Thirst manipulates the addict's vision. The eyes will hunt as much as they gather.

When QAnon launched, Alex called it a fraud. He then tried to launch his own version. The alchemical President ran a double-sided narrative. Trump invites Scott Adams to the White House on Aug 3rd, 2018 and Lionel Media three weeks later. These men are polar opposites on the legitimacy of QAnon. A year prior, Steve Bannon and Sebastian Gorka were banished from the White House weeks before QAnon launched fulltime on the website 4chan. These men are propaganda assets. They are embedded in the media to serve the administration. Assets are liked or hated depending on the operation. They are polarizers used to focus group thought. Assets can be called in for a palette cleanse between scenes like a Scaramucci or an Omarosa. Assets need attention not approval. The emotional charge we get from them is a reflection of their longevity. High-quality assets, positive or negative, are used by puppeteers to manipulate our minds. This happens in politics, advertising, professional sports, and the free market.

Trump is Hulk-a-mania. His fans want him to win. They cheer for him as a byproduct of a reaction to Clinton and Jim Acosta. Acosta is also Trump's asset. Acosta is leveraged by Trump to keep our attention. This is a symbiotic relationship. Acosta is legitimized. CNN remains relevant. CNN is a tool of Trump. Calling Trump the impossible underdog is what gets him labeled the outsider. This is archetypal branding. All male politicians wear make-up and Donald Trump's face is painted orange to align him with the archetype of gold.

Trump harnessed the asset Kanye West when he visited the White House. Kayne publicly fawned over Trump for 48-hours until his lease expired. People forget politics is emotional professional wrestling. The players have to be dramatic and predictable for it to be captivating. Otherwise, people might try to govern themselves. Under mind control, we think it's a good idea if 250 million of us pick one of two colors and whichever psychopath controls that color gets to select who's in charge of the environment, our food, and all the drugs.

Audience engagement is a measurement of obedience. Engagement requires a simplicity found in two political parties. It's why war has two sides. It's why every football game has two teams. Politics is the wrestling for attention. What's beyond the ring?

Trump needs Greater Israel. His bloodline depends on it. If Trump can't stop a false flag in Parkland or Vegas, he's not going to stop a bigger one on the Temple Mount. Once that happens, they won't need to disarm gun owners. The voter is the power behind Trump and right now that power is in the hands of Zionists and dual-citizens. Trump can't be the voice of truth. He can't afford to remember the USS Liberty. He can't afford to say 9/11 was an inside job. Trump has invested his future and his children in Zionism. It's the only way he could secure the job.

Alex Jones recently told Joe Rogan truthers are paranoid schizophrenics. He would go on to say anyone concerned about Zionism, Mossad, or Israel is mentally ill. Jones corrupts the truth simply by standing next to it. The issues he claims to expose are issues he toxically discredits. Alex doesn't spread the truth, he taints it. He reduces it to a cartoon we dismiss as entertainment. Whether or not Alex Jones is even aware of it, he is a spineless tool for Mossad.

"Listen to everyone, read everything, believe nothing unless you can prove it in your own research." – Bill Cooper

Trump's goals are the same as QAnon's. Trump will manipulate the American mind to win. This is what makes him an effective leader.

Saudi Arabi will be blamed for 9/11. Iran will be liberated as we did in Libya. This will lead to the Greater Israel project and a new peace in the Middle East. This phase has been planned for over a century. It is the death of the phoenix with a rising into the eye of a new pyramid. I am not telling you Mossad works against or even with the US government. I am telling you Mossad is the US government. America is vital muscle inside the body of the leviathan. We will bring peace to the Middle East like the mafia brought peace to shopkeepers who paid for protection. Peace is the commodity of bullies. Americans would rather be the New World Order than admit they're the New World Order.

Let America be the emotional litmus test for truthers. There is no joy in swallowing the truth of Zionism. When you do, it answers everything. Zionism is the purpose of the Balfour Declaration. Zionism has embedded itself in the mind of America. It's the sole purpose of Hollywood. It's why Trump looks like the underdog no one expected to win.

If you feel a strong reaction to a political celebrity the system is working. Trump needs an anti-Q task force like Scott Adams, Steve Bannon, Sebastian Gorka, and Jack Posobiec to play the bad wrestlers. An effective psyop injects a synthesis to create anti-synthesis. This is the same mechanics of Hollywood. They tell you what to dislike as often as they tell you what to like. The mind is moved by repulsion as much as it is by attraction. Posobiec's smirky face is effective because it's unsettling. Strong emotions are the fuel of every operation and Jack is one of many employed in the media by Naval Warfare Intelligence. OAN (One America News Network) and QAnon are so similar because they were deployed by the same military task force. The NDAA makes domestic weaponized propaganda legal.

America legally became a propaganda outlet in June of 1942 with the creation of the Office of War Information. Six years later, the OWI turned their focus domestically in the Smith–Mundt Act. We legally pay our government to lie to us. A successful government projects the illusion of total dominance. A world government is required for total control. Behold the Pale Horse of truth. America is the New World Order and Greater Israel is the next incarnation. Can you smell what the Temple Rock is cooking?

CHAPTER EIGHT
Blackmail and Whitemail

Hard work these days is finding an effective way to tell someone something they don't want to hear. Some question why we would even bother. It's a good question. Who wants to ruin someone's day? Even the answer is selfish. We do it for our sanity. We need the environment to reflect our understanding, or else it causes pain. An allergy for dissonance has developed. The years of pretending the insides and outsides could disagree are over. The game of pretending is insulting to everyone involved. Still, many need heroes more than they need the truth.

A belief movement exists today inside the truth movement. Many are angered when you point this out. You can see why they constrict the truth. They are doing it to survive. People will tear you apart for telling them what they don't want to hear. These people don't want truth as much as they want a narrative resolved. But if narratives were true, we wouldn't need to call them narratives.

The psychological narrative has an opening, a middle, and an ending. The timing of these events is designed for engagement and the injection of information. The first rule of theater is never break the fourth wall. It reminds the audience we're only pretending. On April 4th, Chelsea Manning was released from solitary confinement. Two hours later, WikiLeaks tweets that Julian Assange's arrest is imminent. Julian spent seven years locked inside the Ecuadoran embassy while pursuing a romance with Pamela Anderson for the media. Through high-profile media interviews, to speeches from a sovereign porch where he threatened to release a thumb drive, Julian's life has been an archetype of truth on a pedestal. His Hollywood movie plays well because people want to believe the narrative. Belief is the nectar of man. Government needs that honey. More than taxes. More than compliance. Government needs you to believe. It will use dissonance to prove it. Government is a belief the world can be fair if we trust in our authority.

People who know we used energy weapons to melt Paradise, California cling to the idea Julian Assange has thwarted a superpower's best efforts to destroy him. We pretend countries are real when they're lines on a map.

Meanwhile, Bradley Manning, after being tortured in solitary confinement, emerges with a sex change, a fashion spread, and a full pardon from Barack Obama. Manning's gender change was trauma programming. These people squeeze as much juice as they can out of a narrative. Manning went from military torture and suicide watch in November to the cover of Vogue Magazine and a political activism career in August. Our government knows how to program its citizens. Archetypes like Manning, Snowden, and Assange are keys to mind control.

Our world is run by the manipulation of prana. Our fear, anger, and passion fuel a machine that controls us through the art of triangulation. When presented with two choices, the mind will move to the option that causes the least amount of discomfort or calories. Dissonance burns calories. Deciding who's lying becomes an endless exercise in scrutiny. It's a taxing and painful process, which is why so many step off the bus. We've been trained to resolve decisions in thirty-minute sitcoms where the bad guys dress in black, and the good guys always tell the truth. Every solution fits within our budget of attention. Programmers count on the atrophy of attention. Present the flock with two arguments, and you will always have a predictable outcome. The Hegelian dialectic is how they manipulate the world. Thesis and anti-thesis always render synthesis. People who don't understand mind control never look at the final solution. What was the end belief? Who benefited the most? What was left behind? Understanding the technology of narratives starts with understanding oneself. Your endocrine system drives most of your decisions. When you see belief flow through your own body, you understand mind control.

Truth is an emotional journey. We converse with people who

The Technology of Belief

make us feel hollow sometimes. We sense their evacuation as they lie to themselves about what is happening in their world. Consider your own vibration as a kind of homework. You will discover that reason is rarely the conductor. If everyone could handle the truth, lies would die of thirst. Instead, all of us are drunk on propaganda. It pours from clouds so fast we'll never keep up. This is the purpose of Reuters and the Associated Press. They waterboard the mind in the endless barrage of fermented drama. True or false – it doesn't matter. The only requirement is endless.

Our world is not broken. Discerning truth has a spiritual purpose. It reminds us the answers are found inside. Let the storm of doubt make your resolve even stronger. The key is to understand most of your questions are simply self-doubt projected into the world. Lies are like rust in our hearts. God needs each of us to become stainless. We are inside his alchemical tumbler turning ourselves from lead into gold. This gives evil a very specific purpose. God uses it as a kind of whisk frothing us until we rise to the occasion. He's waiting for us to claim sovereignty. We do this by claiming ownership of every photon we give and receive. Energetic consent is the whore of Babylon. God requires us to become pristine.

There is an orchestrated rhythm to the news. The Meuller Report closed down in perfect time for a scene change. Julian Assange, the stubborn sardine, is finally pulled from the London embassy. The Police allow him to clutch a book by Gore Vidal for the cameras. April 11th is 411, the number you're told to dial for information. The U. S. Department of Justice will be charging Assange with conspiracy. They'll claim he aided Bradley Manning in the theft of classified materials. Julian Assange is the closest thing the people have to someone who speaks the truth. But he'll never tell the world what happened.

"I'm constantly annoyed that people are distracted by false conspiracies such as 9/11." – Julian Assange

At best, Julian will be a martyr. The new religion is government and Julian Assange is the Patron Saint of Truth. Most of the world is controlled by blackmail. A lot of evidence is about to be dropped in a giant pile as big as a house. People won't notice that 15% of it is missing. The 85% will be so juicy and shocking it will take over the senses. We are about to partake in an endless buffet of everything but the main course. Who knocked our buildings down? Who killed the people inside? Who used our military to attack our own country? Julian Assange's published record is flawless. The man is the archetype of the whistle. He is also an asset put into play on the world stage to hide an even bigger story – 9/11 was an inside job. I beg you to pay attention to what doesn't come out as we celebrate what does. If we don't pay attention, we're no different than the Romans appeased by a fresh crop of gladiators.

Ignorance is the heroin that keeps us in bed. Our handlers tell us something attacked us in our sleep. 9/11 holds a truth Trump and WikiLeaks don't want you to know. I search the archives hoping to prove myself wrong. Yes, there is some info about 9/11. But it's parsley on meat. WikiLeaks points us to a foreign country's role or a nuanced revelation about the Port Authority. The stability of government is at stake if the people were to become truly awakened. The American psyche is a Gulliver lashed to the ground by millions of ropes. If it finds out what happened during WWII, JFK, or 9/11, it would finally sit up. It takes a lot of mind control to keep that giant on its back. It takes a worldwide effort. Countries aren't what we think they are. If they were, everyone would know the truth.

"I'm not joking when I refer to our country as the United States of Amnesia." – Gore Vidal

The opposite of blackmail is whitemail. Whitemail is when you give someone credit for something they didn't do and let them feed off the reward. This creates a prana contract. Energy slaves are predictable.

They praised Perseus with gifts long before he left to kill Medusa. If he came home empty-handed, he would be a fraud. They made Obama the President of the United States, knowing he wasn't born in Hawaii. They gave him a Nobel Prize and told him to bomb seven nations or be exposed. Obama was dangling by a birth certificate for eight years. Sheriff Joe Arpaio knew the truth. It's why the Sheriff was such a high-value target to Obama. It's also why Trump spent the capital to give him a full pardon.

The Technology of Belief

"Why is it that no one would touch this. A fake, forged government document in the media?" – Sheriff Joe Arpaio

Trump's moral alignment is, at best chaotic good. He plays to win more than he plays for truth. Trump is a realist. We pretend white hats have to follow the rules. We pretend black hats always lie. We are so misguided by Hollywood and political theater. We don't even want a lawful good leader. We're too tired of losing. Under mind control, victory is always more important than wisdom. Wisdom doesn't reward you with laurels. Wisdom is a burden. The truth carries more weight than all the gold in the world.

Dear reader. I insist you believe in yourself as the primary oracle for change. Divest yourself from the system. Otherwise, we fall for the next act. The displacement of your energy is taking away from your precious ability to transmute the world.

CHAPTER NINE
Alchemy of Airships

Deep breath. I'd like to take you back to that time when you were a child jumping on a trampoline. You're not wearing shoes. The sun has yet to consider going down. Bounce after bounce you feel the meat of your cheeks separate from your teeth. You're a combination of enthusiastic and exhausted. You lie down on your back and look up at the clouds and feel as if the earth could let go of you at any moment. There is a joy surging through you that reaches all the way through time to here and now as you read. This joy is a string you pluck on a harp. Listen to its tone. Deep breath. Picture your ideal self. Daydream that person in an ideal location befitting of your ideal life. Take a deep breath as you would imagine that person would. Feel the warm gush of energy rise with the profound feeling of gratitude. Notice who you have become. The feeling overwhelms your mind and renders it speechless. This same feeling sends tingling pulses down your arms as an orange web of synaptic lightning plants you deeper. You are seated in your pelvis like a throne. Your ideal self is who you are right now. The song of perfection is already playing inside of you. You can turn up the volume if you'd like. You can bring that person to the front on the stage like a solo. Gratitude amplifies the signal. When we resonate self, we pluck a bloodline note. It stretches all the way back to that time on the trampoline when you closed your eyes and wondered who you would be when you grew up. The answer is you.

I can't help but wonder now if we grow up naturally. Compulsory education could change us in ways we currently blame on puberty.

The Technology of Belief

When I was ten I was losing my ability to pretend. I would sulk at the feet of three tulip poplar trees I had planted in the backyard. I never saw them grow if they did. I spoke to them anyways. They seemed bored with me as I grew taller. We are insane wizards as children. We summon dragons by rolling their names off our tongue. We vanquish them in valiant pursuits in the sky or underwater. We turn a glass upside-down in the tub as tiny divers swim in and out of the bell for oxygen. Pretending is an alchemical trance induced by the art of visualization.

The magic of sticks was lost when I was eleven. Adults would bribe me with money to call them twigs and rake them away into bags. I could still pretend with the help of certain magic items. A die-cast jet seemed to yield a trance the longest. I could run maneuvers through a tall pine forest with my mind never leaving the cockpit. One day I noticed Legos fell apart too easily. We all go outside one day and forget how to play. This is a kind of death society doesn't acknowledge. It's a loss we never mourn.

Visualization is a form of belief that entrains our will like a metronome. The longer you hold the vision, the deeper you entrain the body. Time builds rapport. Repetition installs vibration. There are four distinct phases of learning that start and end in the unconscious. Knowledge is mimicked when we reach level three. Knowledge is ingrained when we reach level four. Alchemical visualization is a level four skill.

1. Unconscious ignorance – We see no ignorance.
2. Conscious ignorance – We know our ignorance.
3. Conscious competence – We regurgitate.
4. Unconscious competence – We create independently.

During a staring contest, there is never a doubt who wins. Whoever maintains the frame longest wins reality. Both contestants conjure the same vision of, "I am." Both agree to a single rule, "There can be only one." Staring contests are taxing psychologically because something is happening in the aether. Two wills are created in the same sandbox. Under the rule of one, both wills become psychic muscles burning themselves to fatigue. The illusion of scarcity is a powerful motor. It's the secret to mind control. Scarcity turns the quest for wisdom into a race for victory.

There are airships above the clouds we don't see. Our minds only render what we can fathom. We are crouching inside ourselves as a society. We are too afraid of letting go of our master. We pretend it's him keeping us down. We visualized a machine of scarcity and competition. We all agreed to the rule there could be only one. It's not so much that we lost our magic as it has been twisted and turned against us. We still know how to pretend. We are really good at it. We are pretending to be slaves right now.

CHAPTER TEN
A Smooth Criminal

The human psyche can be split like a log yet still appear solid. The mind can exist in wedged chunks under a single sheath of bark. Michael Jackson was two personalities. His core was built like a walk-in closet. On one side was Peter Pan, the superhero. On the other was the Jackal, a pedophile. The public dismisses the reality of Dissociative Identity Disorder (DID). Even the jury at his trial was kept in the dark. Why would the most expensively prosecuted jury in the country not be told about DID?

"We felt that Michael was still a kid, in a man's body." – Paul Rodriguez, Jury Foreman, 2005

Michael was an innovative magician on stage who violated children sexually in private. Michael was an abuse survivor who didn't make it. Joe Jackson pushed Michael and his siblings into the machine. From the age of four, he was raised to be a thoroughbred performer. Each of his victims had the same deposition, "Michael made me feel special." Michael suffered from the ability to make people feel good. Fame is a spotlight harnessing the power of mass neglect. Calling someone star implies everyone watching is not one. No one survives on neglect. They'll chop themselves to bits first.

Famous people are prey on a pedestal. Famous people are predictable because of fame. Famous people don't have free will. It's hard to fathom until you see through the veil of a Tom Cruise, a Kayne West, or a Conor McGregor. These people are exotic meat machines. They have owners who trade them like race cars. Every moment of a star's life is a commodity. Working animals have staff who tend their mane and keep them in the stable. Some dogs learn to chase the rabbit. Some were born to run. Modern high-end human slavery is best understood through one's dopamine receptors. Man will always serve the chemicals that taint his thinking.

Michael Jackson said about spending time with children, "It's like I see myself in the mirror." He never asked his reflection to change his ways. Michael Jackson wasn't lying. That part of him would never hurt a child. Jackson said, "I would slit my wrist first. I would never do anything like that. That's not Michael Jackson. I'm Sorry. That's someone else." He said later about sharing a bed with children, "So what. Who's Jack the Ripper in the room?"

It's normal to become enamored by the hero alter. That part of the psyche is designed to be magnetic. It will never process accountability across the cracks. Michael named his child, "blanket" to hide it from the Jackal. He'd drop it from a balcony if someone would save it from its father. Michael bleached his skin and cut his nose off like an Egyptian statue to mute his own power. Michael was battling a creature embedded in his bones. Michael didn't hate his Jackal. He lacked the ability to fathom its existence. Michael would have to die two deaths to embrace his shadow. Neverland was a place of fantasy with plenty of doors and gates.

People are under a trance that Jackson is innocent. This is the display of how powerful fame can be. Quincy Jones molded his music. Scorsese directed his videos. Michael's life was a traded pony fully initiated into the system. No one that big in the industry gets to keep their innocence. Parents were tossing their children over the gates of Neverland like fishermen casting lures. They fed him babies like he was Moloch. They gave him the keys to their bloodline in hopes he would turn their offspring. Macaulay Culkin was turned by the Jackal. He fetishizes himself now as an innocent bunny. He misses the days when someone like Michael found him special. These are the days of the voodoo sacrifice. The witch doctor moonwalks across a Pepsi stage wearing a silver glove.

Michael Jackson was traded into slavery. People like Quincy Jones, Steven Spielberg, and Oprah Winfrey are pirates who profited from the transaction. Quincy met Michael Jackson at age 12. Michael would later audition his own entourage of children ages

The Technology of Belief

12-14. A war developed for Michael's leash between the system and Joe Jackson. Manipulation turned Michael and some of his siblings against Joe Jackson. It wasn't that hard. Michael explained his abuse growing up on Oprah's couch. La Toya did the same a few weeks later. The Jackson's were thoroughbreds stolen by the vampires of Hollywood who flaunted pedophilia. Joe Jackson died last Summer with all of his living children by his side. Oprah celebrated a screening of their new documentary *Leaving Neverland* just six months later on a yacht with David Geffen. The most effective lie is the constriction of the truth painted on a silver screen. This is an injection into our mythology. A cementing of a lie. Is it any wonder Michael paid $150,000 to a voodoo priest to slaughter cows in a magic attack against Geffen and Spielberg?

Michael was broken into music industry by pedophilia. They turned him into a smooth criminal with sex parties, love-bombing, and the promises of playing Peter Pan on the silver screen. Steven Spielberg made Michael Jackson feel special. He did the same to Tom Hanks. Slavery is the illusion of a love that returns if we try harder.

Falling for a love bomb requires self-abandonment. The empty watch tower waves no red flag. Michael's first nose job was in 1979 at age 21. In 1983, he did it again. Three years later he did it a third time. In 1988 he cleft his chin. In 1992 he started cutting his eyelids. In 1994, his nose was a sharpened pencil. By 1995, Michael Jackson finally looked like a Jackal. He had over one-hundred professionally-assisted self-mutilations. Hollywood calls these satanic rituals "plastic surgery."

A long time ago, during the days of human sacrifice, natives would cut pieces of their skin off and bleed while the people watched. A golden child would step up on a pedestal and chop off his manhood to cast it into the fire while the crowd cheered. Those times are still here. Doctors must wake up from the spell and see the act of violence. They are trained from circumcision that the practice of mutilation is exempt from moral scrutiny. A child lacks the ability to say no. A mother lacks the ability to ask. Michael didn't know what he was doing. He lacked informed consent.

"I've had no plastic surgery on my face, just my nose. It helped me breathe better so I can hit higher notes. I am telling you the honest

truth, I didn't do anything to my face." - Michael Jackson

The Michael Jackson mutilations were capitalized on by our programmers to taint the archetype of patriarchy. The supremacy of self is constantly shamed in a society under mind control. We insist Bradley Manning, Bruce Jenner, and Michael Jackson is simply exercising their autonomy. How does a slave exercise autonomy?

Human slavery is real. It has nothing to do with chains. In today's Rome, there are slaves for the masses. We call these slaves, "stars." Cultural mind control is done through the scripting of these human archetypes. Campaigns against self-worth can be introduced under the evils of patriarchy. The ego can be beaten into submission with the promise of ascension. Mankind can be steered with the powers of dopamine and its restriction. Groups can be coddled into a warm tub to relax. There is power from the threat of taking the water away. Today's slave isn't chained to a post. Their reigns are loosely draped like a loyal broken horse parked outside a saloon.

CHAPTER ELEVEN
Fire & Isis

In Paris, we read, "The cause of the fire is unknown" and call ourselves informed. We hear from imaginary people on the scene that Notre Dame was an accident. We are content with the official handle on the situation. There was a fire. Its cause is unknown. This is our working narrative. We can link to the story and never be wrong. Anyone who says otherwise is misinformed. News is a television jingle. We sing it to our loved ones and friends. We share it over the water cooler of social media. No one can disagree as long as we mimic it verbatim. Miss one word and the room knows. News is a karaoke machine. Being informed is the same as singing with the prompter. The source of news is a one-way supply of information.

Narratives aren't news as much as they are energy. That energy is the drama of receiving new information. We bestow fame on those who deliver it to us. We reward the first runner from Gaul to bring us news of the Emperor's victory. The second runner is kept waiting outside of court. Both runners deliver the same gift of news. Yet only the first is rewarded. The prize they bring isn't information so much as it is freshness.

In the excitement of fresh ideas, we forget something important. Omissions are facts, too. The space between the stories tells most of the truth. We require seers to look through the obsidian ink and peer beyond the editor's shoulder at what was missing. We must divine our truth. "The cause of the fire is unknown" is not the truth. Nor is it information. It simply puts to rest your desire for an answer.

If Presidential debates were for scrutiny, someone would ask what happened to the USS Liberty. If debates were for ratings, someone would ask what happened to WTC Building 7. If debates were mind control, no one would ask these questions. Under mind control, the cause of the fire will always be unknown.

When I was a boy I walked through the doors of Our Lady of Paris Cathedral. It was darker than I expected from being outside in the bright sun. I was wearing shiny patent-leather shoes. I was draped in a red robe under a white tunic. I was carrying a song from America. There were forty-four boys in the choir. Most of our offerings came in Latin. My favorite was the canticle, "Magnificat." The Song of Mary. We sat in ancient pews on either side of the aisle with our backs pushed against the cold stone walls. When you sing in a cathedral with such a high ceiling you never quite hear your own voice. The internal sky pulls you up in a rapture of dying thirst. Notre Dame was the belly of a giant whale and its windows of purple and blue were frozen mandalas that moo'ed. Notre Dame has always been a weeping mother. She's forever beached on the island of Paris surrounded by the Seine. Before she was draped in Catholicism, this holy ground was the Temple of Isis. In 1163, The land of the Parisii had been taken over again by a new kind of Rome. Notre Dame was a two-hundred-year-long construction. It survived Hitler and two world wars. That we accept the cause of the fire as "unknown" is the real tragedy.

Notre Dame has been attacked several times in the past few years. In the first week of February, five churches were vandalized in the same week. The news says things like, "it is unclear if these incidents are related." In 2016, a priest in Normandy had his throat slit by hostage-takers. The news was very clear on reporting it was jihadists. We know there have been hundreds of Catholic church vandalism in the past few years. We know they have escalated dramatically starting in February.

There is a cause for every fire. Besides jihadists, there are two suspects for this vandalism. The first is the Catholic church. These attacks come when the church is hurting most for sympathy. When Father Roman Galadza was asked about the fire at St. Elias the Prophet in Ukraine he said, "We'll just have to be more careful, I guess." In hundreds of fires, his was one of the few reported as an accident. The next suspect is the cabal moving human cattle. A

The Technology of Belief

well-coordinated climate is important when seeking new pastures. As more YellowVest protest images get published it seems impossible to ignore the amount of propaganda injected. On one hand, the cabal wants unrest. On the other, they want the YellowVest to appear savage and violent to keep the people inside. Their goal is a complacent sense of dread. They want people desperate but still hopeful. Hope was the last evil from Pandora's box before she closed the lid. People are always confused about why someone would call hope evil. Hope is faith giving up of itself. It's saturated with desperation.

Crime pays. False Flags pay more. Parkland paid $11 million. Notre Dame $844 million. 9/11 $2.3 trillion and counting. Our world is full of cowards who get angry when you show them the truth. Cowardice is a trance induced by a prolonged ritual of manners. We challenge who we respect. We push who we think can take it. To see better in someone is the same as calling them out for slouching. I see better in people. I see better in our discernment. Men don't want to be liked as much as heard. So put your ear to the track and hear the vibrations. There were three fires burned in three churches three days before the Passover. New York, Paris, and a mosque in Jerusalem burned under a waxing pink moon that crested early Good Friday morning.

Men with giant spoons stir the world. We sit content in our warm pot smelling each other as they cook us alive. There's a hole in the roof of a Cathedral that's been closed for centuries. The spirit of Isis and the waxing moon touch ground again for the very first time. All of this during Holy Week. The house of Isis is open now and the light from the crescent has come home. The third temple of Jerusalem was opened this year with the burning of incense and the blood of an innocent lamb. The hundred-year plan at Balfour is taking its seat on the stage for a new religion.

They removed the heads of every Saint before the statues were raptured to heaven by helicopter. Photos show where the arms on one are higher than the head. The staff is higher than the shoulders on another. This wasn't about clearance. These statues were decapitated with a blowtorch and airlifted on video in broad daylight. None of them were covered like a mover might wrap your own statue from your home. They were headless and on display for a reason. This was days before the fire. This question is missing from the agenda. The New York Times describes the world's most important church with the world's most advanced fire suppression system an old dusty barn that was built to burn. This directly contradicts French Former Chief Architect Benjamin Mounton who oversaw extensive electrical and fire suppression upgrades in 2010 saying, "You have two men who are there 24/7 to go up and check any warning." There are many more questions missing on purpose. We must learn to see through the cracks to render the truth.

Considering the attack in Sri Lanka over the weekend with the emerging violence across Europe, it's accurate to say it was a "terrorist" attack that was semi-thwarted. Officials are calling it an accident to mute the rippling effects. Professional instigators have produced another global war between the Cross and the Crescent, Jesus vs Mohammed, in the colosseum. An important mosque as symbolic of Notre Dame will be next. There are three children of Abraham. Two of them are being pushed into the ring for battle.

CHAPTER TWELVE
The Satanic Messiah

There are three children of Abraham. Each worships the same god but argue about its representative. Each child reads from a different book that claims to be the truth. All three children believe in one god but disagree about its messenger. The disagreements have grown so deep we call them religion. This is what happens when you tell people there can be only one. Monotheism is the belief in a number. Scripture is the belief in a word. Neither of these are belief in God.

The three children of Abraham are Christianity, Islam, and Judaism. Each child calls God a different name like Yahweh, Jehovah, or Allah. The god of the Old Testament is the same god of the Quran. Chronologically, the prophets of God have been Adam, Abraham, Moses, Jesus, and Muhammad. With every new messiah, a new branch springs from the trunk. Judaism's Messiah was Sabbatai Zevi. Jewish scholars dismiss him now as a pseudo-messiah with mental problems, but his profound effect on modern life is ignored. Zevi was a dark crystal. A first resonator of evil. He installed ritual depravity and ancestral pedophilia into the population. Half of all Jews believed Zevi was their Messiah. His inversion of Talmudic law was the bloodroot of modern Satanism and underground worship.

Sabbatai Zevi was born in Turkey on the 9th of Av in 1626. He was named for the planet Saturn. He excelled at a young age as a kabbalist and mystic. At age 22, he declared himself the Messianic Redeemer predicted in the Zohar. Zevi was a man of deep conviction who detested boundaries. He recited the sacred name for God to a crowd who gasped at his blasphemy. He was a liberator of decency. He awakened himself and others with the raping of taboo. Zevi claimed he could levitate, but only the worthy could see it. His behavior and passion developed a following in the town of Smyrna until he was banished in 1651. There is much evidence Sabbatai would be labeled bipolar.

We call people bipolar for possessing a deep wave. Their rhythm carries too much amplitude for comfort. Society accepts people who are predictable and operate within a threshold. You can only be so short or so tall. Someone who is bipolar violates these requirements and suffers the consequences. Society governs the throttle of chemicals inside each of us using the sociological tools of shame and fame. Zevi's condition was groomed not born. If he was too sad, he was shunned. If he was enthusiastic, he was rewarded. Zevi was charismatic, impressive and brilliant when he was high. He was dark, destructive and dangerous when he was low. This made Zevi a magnetic and reluctant messiah.

As Zevi's fame grew his mind split like a harlequin. It was all he could do to cling to God. In 1658, he declared himself wed to the Torah in a public ceremony. This act drew the attention of a gifted alchemist and orator, Abraham Yachini. Yachini understood Zevi's nature and used him as a tool. Yachini was a belief technologist.

Yachini asked Zevi into his home, "I have an important scripture we must read together in private. It concerns your destiny." Zevi joined Yachini on the same cushion at his table. Yachini unrolled the long scroll to introduce it, "The Great Wisdom of Solomon is written in an archaic language very few know. These are the words of Abraham who spoke of what's to come. I assure you Zevi, I have seen many things written come true." Zevi nodded in reverence at the privilege of witness. Yachini began his incantation with a firm voice.

"A son will be born in the year 5386 [1626 AD] to Mordecai Ẓebi, and he will be called Sabbatai. He will humble the great dragon. He, the true Messiah, will sit upon God's throne."

Yachini watched Zevi's eyes heave and said, "Was your father's name, Mordecai?" Zevi nodded. "And you were born in this year?" Zevi nodded again. "Do you understand your destiny?" Zevi nodded slowly a final time as tears kicked like frogs from his cheeks. The chemical shower drenched his brain like a sauna. His whole life, Zevi felt he was misunderstood. Yachini brought his isolation to the surface like a tilled garden.

The power of one's belief is multiplied exponentially by the words of another. Yachini had penned the passage for Zevi four days prior with details divulged from his crew. The scroll was unrolled as if the prophecy had been written centuries before. He would lie to Zevi and call his actions righteous. He was adjusting Zevi's spine to tune a higher signal. Truth is a mushroom that needs the sound of thunder to drop its spores. Zevi must be thoroughly seated in his own confidence to serve the world.

Zevi accepted his divine crown as the dopamine flooded his brain with gratitude. Zevi was both liberated and enslaved by Yachini's magic. This is how you catch a rising star. Zevi was a golden child.

The Torah has 613 commandments, and Zevi would say all of them blocked divinity's flow. Passion was Zevi's umbilical to God. If the cord was restricted, Zevi would suffocate in darkness. People expected a gurgling miracle from his lips and abandoned him the moment he spat dust. This pushed Zevi into higher states of depravity. In Constantinople, he convinced a husband to mount his wife on a tavern table while everyone watched. Over the years, this act devolved into ritual urination. Sabbatai's bipolar crashes made him best suited as a traveling mystic. He left a huge impression wherever he went and a big hole once he was gone.

He was two years in Cairo before meeting Nathan of Gaza. Zevi wanted help for the vacuous depressions. Nathan of Gaza was locally known for his sacred vision of Ezekial's chariot during a week-long fast. Nathan was as radical as Zevi but for different reasons. In his work, A Treatise on Dragons, he called for the complete elimination of Torah Law. Nathan supported Zevi's claim as the Jewish Messiah but ignored his battle with depression. To him, Zevi was gifted not cursed. He lifted Zevi up high on the pedestal as the divine golden child.

Half of the world's Jewish population believed Sabbatai Zevi was the Messiah. One million followers saw him as an archetype of God. His acts of liberation and depravity trauma programmed a population with shame and adrenaline. A major cleft in the branch of Judaism was cut and Satanism was born.

"Blessed be he who permits the forbidden." – Sabbatai Zevi

Morality comes with its own virginity. There are lines one cannot uncross. Exposing the ears to a booming sound destroys their fidelity for softness. The same fate awaits a morality that's been pulverized. Those limits once crossed don't grow back. Evil only rewards for advancement. This forces followers to push more every time. Depravity becomes a chemical addiction. Zevi turned half the Jewish population into addicts.

The golden child we see in the spotlight is rarely in control. They become a slave to the desire of their followers. Nathan of Gaza controlled Zevi with a new prophecy, "The Messiah (Zevi) would lead the Ten Lost Tribes back to the Holy Land, riding on a lion with a seven-headed dragon in its jaws." On Sept 16th, 1666, Zevi was given an audience with the Sultan. Zevi demanded Jerusalem on behalf of his people. The Sultan answered Zevi with two choices. Zevi could suffer a volley of arrows and let God protect him or he could convert to Islam. Levi converted to Islam and the crypto-Jew was born.

The followers of Zevi were devastated. Many had defiled their family, their faith, and their holy laws for a man who abandoned them. Some removed the roof from their home trusting Zevi's words that God would rapture them to Jerusalem. In one generation, millions of Jews were installed with Zevi's bipolar possession. Abandonment is an echo of depravity. It's implanted in our DNA for generations to overcome. Sabbateanism is still tremoring today. It hides in bloodlines behind the vestments of many religions. Each secretly believing the more you defile, the sooner God will come. These beliefs don't have to exist consciously to be followed. A belief can be planted deep in the DNA through trauma. It activates itself like a terror cell with a phone call. These people actively spit in the face of God for his attention. They run our media, our corporations, our governments, and our wars. They are the serpent cult of Catholicism, the crypto-Jews of Islam, the Satanists of Hollywood, the Mossad of Israel and everyone in the worldwide cabal. They are the living reverberation of evil on a quest of assimilation. The fact we don't have a common name for them shows how ruthless and effective they truly are. They're not Jews – they're Saturnians. They are the children of Saturn. In 1666 they inverted the Torah. Today, they openly celebrate a burnt offering called the Holocaust. They celebrate the fire of 6 million and use it as a torch. This mass sacrifice is payment for the delivery of Judgement Day. These people are convinced God will come if they sin loud enough.

CHAPTER THIRTEEN
The Trojan Horse of Zionism

Zionism is a Death Star orbiting the forest moon of America. The empire calls its citizens anti-semitic monkey-people tainted with measles. We are told the rebels must be vaccinated from their own disease. Our madness is chained under the master's desk. We smell the cigar smoke of Zionism wafting. Mossad has Trump's head in its mouth. America First wants him to step on the snake while Zionists are offering a saddle. America loves Zionism more than it loves Israel. We are programmed by a hypnotic number six million. If World War II was about the extermination of one race why did he start with the French? Why did Hitler invade Russia? *Mein Kampf* reveals a disdain for subterranean nations living inside of what Hitler called the Fatherland. Hitler's first beef wasn't with the Jews. The mind control is so bad I have trouble even typing the word, "Hitler." As if reading the amended plaques at Auschwitz makes me a bad person. Or consulting the world census makes me a perpetrator of genocide. We are punished and shamed for snooping into "settled" history. My recent article on Judaism was flagged by social media. I have been suspended from my third platform. The most effective mind control is a tiny gnat landing in the eye. It teaches you to graze where the bugs won't bother. One tiny kamikaze insect is all it takes to move a giant.

If an ideology exploits the suffering of its people then that suffering becomes a harvest. Could this be why it's called a burnt offering instead of a murder? No one honors the dead by exaggerating. Would we remember 9/11 more or less by curbing the truth or fudging the numbers? We believe in the chosen people because we would burn in hell for doubting it. We are too afraid to even ask for identification. The biblical chosen people could be Ethiopians or even Irish. The belief in a Zionist state is religious narcissism. The only cure is to stop giving it our supply. American Zionism is the enabler. Antisemitism is the flying monkey.

"A Jew is not a Zionist" – *Rabbi Dovid Weiss*

There is no difference between globalism and Greater Israel. Zionism declares every non-Jew in the world unchosen. Every Abrahamic religion has been told their book is sacred and true. If religions were planets, Christianity would be the sun and Islam, the moon. Judaism would be Saturn. Abraham's hands turn the blades of a cosmic great work. It gives us the golden child of Christianity and the scapegoat of Islam. Both are manipulated by the hands of Saturn. In abuse, the golden child is abandoned for their failures while the scapegoat is neglected for their triumphs. Only the children of Saturn benefit from the sacrifice of others. Julius Caesar was the messiah of Rome. Sabbatai Zevi was the messiah for Judaism. Hitler was the messiah for Germany. We are programmed to follow a messiah. We are programmed to seek deliverance. We are programmed to think like slaves.

We are born into a slow cyclone. Modern interferometry suggests the earth is stationary as the aether slowly rotates around us. God is stirring our magic potion for a reason. Once you know the recipe, the violence becomes predictable. Mankind is yeast rising up from his knees in the heat. Human slavery is a kind of hatchling. We were enslaved to each other long before we were enslaved to Rome. Liberation is a blossom that keeps unfolding. Our thirst stretches the tongue from its cave to taste rain.

This week, Florida unanimously passed public school anti-semitic legislation making it illegal to criticize the Israeli people. Veterans who said Israel goaded America into war with Egypt by attacking the USS Liberty can go to jail for their testimony. Every hour another veteran commits suicide and the government assures us none of them will be anti-semitic. We are creating yet another special class of citizen who will suffer on a golden pedestal. Anti-hate laws are the Trojan horse we wheeled into our city while marveling at how virtuous we must be for taking the gift. It will be night soon. The wooden belly of the beast will crack open and men will descend from ropes to slice open our throats. They have come for our voice box. I ask what's the difference between "Trust the plan" and Stockholm syndrome. The answers aren't satisfying.

Wikipedia says, "Stockholm syndrome is a condition that causes hostages to develop a psychological alliance with their captors as a survival strategy." When Patty Hearst was taken hostage at age 20 by the SLA she believed her captors were compassionate. She hung hope on the notion they would be moved by her pain. She pretended a kind heart was inside their chest so she had something to reach. Nineteen months later, in 1974, Patty was helping her captors rob the Hibernia Bank. A month later, Patty had renamed herself Tania and emptied an automatic weapon into a storefront during a getaway. Patty Hearst was convicted for armed robbery and pardoned by Bill Clinton. The inventor of yellow journalism's granddaughter was shown mercy for multiple felonies.

The FBI's COINTELPRO program was installing domestic terrorist programs like the SLA nationwide from Berkeley to Mississippi. The Weather Underground and the Congress of Racial Equality (CORE) were domestic training cells posing as anti-racist/feminist/communist organizations. They recruited radicals from inside prisons and used media to arrange false flags attacks in Mississippi and Kent State. Mind control is the selection of gladiators for the Roman Coliseum. The civil rights movement we study today was installed by federal gangsters cashing in on domestic strife.

During JFK and 9/11 our nation was taken hostage by voodoo witch doctors blowing dust in our face. We react to the spell in a predictable way because we are well trained. A cobra with seven heads has been charming us from the beginning. Each strike of its mouth comes from a new direction. We do well to keep our feet off the ground and dance. The right caters to Zionism because they hate pedophilia and corruption. The left caters to Zionism because they fear their own racism and hate to see suffering. We could point our weapons at the real Pharaoh anytime now. He is on the hill fawned over by our children while we tend to his land with our blood. We are epigenetically conditioned to serve the chosen. This is the technology of belief.

The Technology of Belief

Most of the nation trusts the government. Most of its citizens mistrust big business. Both seem to forget that the government is a corporation. We trust our flag more than its people. We trust the stories of Paul Revere we heard as children. We trust we have the moxie to dump tea or to muster a militia in a dawn's early light. We extend our trust to the government like a free handshake. We believe our shaking makes us more civilized. This belief is how you break a horse. The owner sticks his fingers in our mouth to prove we won't bite.

It is illegal in Sweden and Germany to homeschool. This raises an important question. What's the difference between a child detention facility and compulsory government-induced education? In France, it's illegal to even question the purpose or motivation of the Holocaust. The definition of Holocaust is burnt offering. The six million number has been pushed in newspapers as early as 1915. Six is the atomic number for carbon. Carbon is what's left after a burnt offering. There is a single evil eye hiding behind Atheism, Zionism, Judaism, Catholicism, and Christianity. It wants to send a message to God. It's doing so through symbolism and human sacrifice. The legs and arms of this creature are American Zionism.

The brainwashing is worldwide and fulltime. There are not enough hours in the day to undo the empire's dissonance. Europe is drowning in quicksand. America is up to its knees in the rising tide as we adjust to the smell. People don't believe in humans anymore, they'd rather wait for a hero. This is what happens when you implant trauma from a seven-headed cobra. Behold the Trojan Horse of Zionism. The true history of the second world war masks a century of globalism creeping. Nationalism is attacked because it's the only thing standing in its way. Now we hear being anti-zionist is anti-semitic. This is the perpetrator hiding behind the victim.

"May it please you to prosper Zion, to build up the walls of Jerusalem. Then you will delight in the sacrifices of the righteous, in burnt offerings offered whole; then bulls will be offered on your altar." - Psalm 51:18-19

German historian David Irving has been imprisoned multiple times for researching his book, *Hitler's War*. He has been removed from airplanes, denied at border crossings, and threatened with prison for researching world history. Every day, more of his videos are removed from the internet and banned from our schools. No one disputes Irving's work. The Auschwitz Museum in Germany has verified his research for the British courts. Zionists don't want the real story of World War II coming out. This is why they ended free speech in Florida.

The difficulty in speaking about a topic is directly proportional to its propaganda. Asking, "Is that true?" honors our history and never disparages it. Questions are the focus of our attention which is all a human truly is – conscious attention. Historical fascists insist we must not invest the effort to re-examine. They want the cake they were fed in school to keep. Psychopaths treat their victims like sandbags. Psychopathy has dictated every page of the history book. It is healthy to start from a place of doubt and work our way out. Blaming nationalism for genocide is a cover. Use this as a cipher. It's a clue to notice just how much effort is spent trying to shame independence. Ideas like nationalism are deadly to a global crime syndicate. America is an Achilles' heel. This is why the American mind was subdued by a media cabal after World War II. They need us to abandon the concept of sovereignty. It's confusing to hold truth now. They made it that way on purpose. It requires calories every day to withstand a storm of fairytales about a war between good and evil. It looks better in print than saying the war was the manipulated vs the goaded.

Asking if Israel has the right to exist is the same as asking if they have a right to Palestinian genocide. When we believe in chosen people we give them consent to remove the unchosen. We trust Zionists and their plans for America. We know not what we do. Congress has no war powers which means the people have no say. Florida has lost the first amendment in a unanimous vote and no one is complaining. Don't forget what has happened to us. We rode shotgun in a van for eight years while our captor bombed seven nations with a Nobel Peace Prize for a hood ornament. That maniacal laugh you hear in the back is from people we pay with our attention and taxes. All of the guns are loaded for the next heist. The Third Temple and Iran are just around the corner. The Zionist plan has been in place since the Balfour Declaration. At this point, insanity is the only thing keeping us in the car.

CHAPTER FOURTEEN
The SDK of Magic

You are reading a page from a manual of lost technology. It works like Amazon Alexa but far more advanced. This technology doesn't come in a box or require a monthly subscription. It listens for your commands right now as you read. This technology draws its power from the ground and atmosphere and tunes into your body's antenna when you call it. This technology is belief.

The word Abracadabra is Hebrew for "I will create as I speak." In Aramaic, it means "I create like the word." Magic is a program launched from the voice box written in the language of sound. You are born speaking a dialect of music. Your gasps, howls, grunts, and sighs tell more than letters ever could. We are singing reeds from the womb casting the spell of "I am." Words like sand crumble in comprehension while our tone is a sturdy damp clay. We are castles of ego gloriously dreaming in the face of the tide. Our words say what we feel not what we think. Pluck will's cord from deep in the belly. Give a boom to the voice that comes from your bones. You are an ancestral drum telling the future what will come. Your bloodline is the sacred spark in a quest for fire.

You are the alchemist. You take an oath that your tongue may only speak the truth. Sarcasm makes you a cackling hyena. You are a pharaoh who rises the sun with your palm. You are the message, not its reaction. Strike tones in the matrix with the power of your cords. The two towers in your throat are a magi's lens. Vibration bears a fruit of many colors. This is the body's aura beaming. Find yourself worthy of its coat and you will learn to glow. There is too much darkness in this world to not be a rainbow. Aura lives in the eye of the beholder. It is the emanation of someone's truth. This is the harmonic manifestation of logos.

Magic is an ancient API with a simple protocol. Magic hears your will and holds it accountable for its commandments. The user launches a new program by casting a circle. This is your holodeck in the aether. A circle is as easy as sweeping a broom to form a boundary. You clean a space to welcome something new. This makes a landing beacon for its arrival. A circle can be of any size or shape. Claiming space is how the alchemist draws power. The island of Manhattan was a ritual circle of water on 9/11. The homeowner in the suburbs mows a ritual circle of grass in the summer. Once inside the circle, the alchemist becomes a four-dimensional wand. Power is lost and gained depending on the desire. A man's garage is a magic circle. A yoga mat is a magic circle. A cold shower is a circle. The cab in a carpenter's truck is a circle. That place in front of the kitchen sink is a circle. A bathroom stall when someone's drunk is a circle. The bed in your bedroom, your mother's arms, your face buried in kittens. These spaces are circles where we power up. Energy comes from remembering you are the most powerful person in the world.

Magic props are furniture on the set of a movie. The purpose of the prop is to convey the alchemist's intent to the aether, his audience. A prop represents a magical program you want to launch in the ritual. The ritual of lighting sage launches a cleansing program. It requires no utterance since it has been symbolically endowed. If props give you belief you must use them. Belief is the motor of will. If a prop is contrived it will fail. This is why symbology is so powerful. As a meta-language, symbols have been charged for centuries by alchemists. Knowing their history enriches the quality of belief. Magic is a flow state of belief. You feed it by tapping into the archetypes. A 60-year-old woman in St Louis believes she is the reincarnation of Cleopatra. She carries herself proudly through the mall in leopard pants because of it. She is a walking goddess in her belief. That energy comes through the archetype of the Queen of the Nile. The power of masks, props, or uniforms plugs us into our archetypes. We stand up straight in their current. They belong to each of us. Independently. They are the stilts we use to shed what's meek and stretch our wings.

The Technology of Belief

Both fear and confidence strengthen our beliefs. My cat was an acrobatic ninja on our balcony. He would jump on to the handrail towering three stories over a cliff. His acrobat sent a gush of adrenaline through my shoulders and knees. I pulled back inside and let him have the porch all to itself. His belief gave him power. My belief took power away. Belief is vital to a trance and everything we do is from belief.

Prana is life-force. This can be intimidating to the starved. Enthusiasm is a measure of prana. When lost or stolen, pay attention to how it happened. The world is full of vampires cultivating nectar for slaughter. Be sure the draining of your power has your full consent. You are responsible for where it goes and what it's used for. Consent is a vow of commitment for the alchemist.

Wishes are thought. Spells are spoken. How many of us voice our will aloud to the world? The language of magic is an utterance. Words are as sacred as we make them. Build up prana in the trachea by what you speak. Words are hieroglyphic sigils holding a power to unlock chemicals in a listener's brain. Like the biology of amino acids, language becomes a chemical sandwich. Spells are meals served in courses throughout the life of an incantation. Tell life what it must give you to keep you satisfied. Make every molecule believe you mean business. This is what we do when we save a life, settle an argument, sell a house, or escape a speeding ticket. A conversation is the secretion of chemicals inducing a mutual trance.

Before you can cast a spell you should be sure you're not living in someone's trance. Begin by asking yourself, "Am I sovereign?" Talk to your body. Allow it time to give you an answer. By speaking to the body out loud you open a direct connection to your mitochondria. Your body is giving you constant feedback. Listen to your reflection in the mirror. The pain in your neck. The warm glow from your chest. Feel it from the ends of your skin. Are you sovereign? If you feel hesitation, ask again. If the answer is "no" you have the power to fix it. Instantly tell your body you hold the reins. It will welcome you as captain. If you feel under attack or out of control you can employ an archetype. Jesus is a power station of ancient technology. His archetype has an aura that you can access 24/7. Every prayer activates the built-in recovery program to protect you while your system is restored. This auto-pilot comes with a subscription to the Holy Ghost. You can take back the wheel whenever you are ready.

Make the language of your spells your own. The vibration of magic is more important than its syllables. Be sure a confident wind rolls through the twin towers of your throat. Study conscious language to improve your spell's energy, frequency, and vibration. Remember that words aren't spells. It's what the vocal cords do with words that make them powerful. Your imagination is power. You play words in your head when you read. You are doing it now. Your vibration is the magic, not the words. Sound your will to the world. Dictate its delivery schedule. Command the aether like Zeus from the atop Mount Olympus. This is the meaning of so mote it be.

Alchemists step out of compulsory education dripping in mind control. Many spend their lives in a trance about currency. We've been taught from our youth our desire for money is evil or repulsive. We've been taught our own time is measured in dollars. This is a trance. If you've struggled with money use this spell to break free.

Step 1: The Bible doesn't say money is the root of all evil. It says, "For the love of money is the root of all evil." Having wealth is not evil. Sweep that mind control out of your circle.

Step 2: Money is an energy current. Currency is a two-way street. Feel worthy of your own value. You have the ability to output fifty amps but your psychology only thinks you're worthy of three. Stop cheating the world of your potential. You're not being humble by being meek. You're being selfish with the resistance.

Step 3: Shock the world with current. Output who you are from the ground of your body and voice. Erupt like a volcano of plasma. God trusts the potential he gave you. There are no mistakes from his factory. Your capabilities are warranted and you are being held accountable for how you use it. The stage is yours, you are wasting time wondering if you should be up here.

CHAPTER FIFTEEN
The Man from Katuah

I am a man from Katuah, the Blue Ridge bioregion of the southern Appalachians. I am the living soil. I am land incarnate. I am the vine, the flower, and the thorn. I am its airspace and water. I am its militia. I am its fingers, ears, and voice. During the great trance, this geography was called North Carolina. We don't have states anymore. We live in a confederacy of bioregions. North Carolina dissolved into four parts and seceded from the union in the year 2020. It was the year of American independence. Maine led the charge on July 4th to a nationwide great awakening. This was not a revolution. It was an exodus from slavery.

The exodus was broadcast on social media. Millions of Americans withdrew their consent from the District of Columbia through a personal declaration of independence. Sovereignty pulled the sword from the stone and man became king. The declaration was simple:

"This is my Declaration of Independence. I, [Your Name], void all contracts made under the person, [BIRTH NAME IN ALL CAPS], with the corporation The United States of America. I am a natural born freeman of the land. I am not a corporate fiction. I am the living soil."

The capital of my bioregion is Qualla, the former territory of the Eastern Cherokee Nation in western North Carolina. Katuah and the Cherokee Nation merged in 2021. This was how Katuah got its name. Katuah is classified a class III ecoregion of the United States. As a sovereign of Katuah, I attend the local council held each waxing moon on Grandfather Mountain. The larger regional pow wow in Qualla happens every summer on the solstice. We have no leaders – only agreements. Government is a tapestry of agreements.

I came to Qualla bearing five bushels of fresh apples and four hundred dollars. I set up camp and loaded my bicycle like a burro to transport them to the giving tent. Once there, my cash gift was converted into Ethereum, the digital currency of Katuah. They asked me to take the apples to Sarah in the catering tent who gave them to Dave at the juice bar. He mentioned he might turn some of them into pies. It brought me joy to see the value of my offering. It made me want to bring more next time. Contributions are uplifting. Taxation is deflating. The energy we put into government is the energy that is returned.

I parked my bicycle near a hitching post of horses. I tipped my hat to them and sauntered through the fairgrounds to settle at a picnic table. I launched the Katuah app on my phone and saw my cash offering appear in my balance. Using my finger, I dragged most of my gift into capital improvements for the new water tower. The rest went to the new dock I wanted to see at Buckeye Lake. I clicked the filters tab and added a boycott rule for the proposed asphalt plant. I scanned the onscreen map to find the natural resources tent. I planned to see the presentation on the Great Chestnut Tree project. Last year, this program won funding. Enough people had voted with their fingers and the project was born. I might give more of my current this year depending on their progress. Current is what we call bioregional money. It makes it easy to distinguish from fiat cash. I can change who and what gets my current as often as I like. I can adjust it throughout the year as my position evolves. This is the beauty of bioregionalism on a blockchain. The entire government is an open-source white paper.

The rattles of dissonance fall limp when the tail is cut.

Before the rise of Katuah, our country was indentured under Admiralty law. Black wizards claimed our souls at birth and the people felt unworthy to stop them. We were born as meat contracts with 40% of our labor extracted under the threat of violence. Taxation is theft but no one seemed to mind. We were sharecroppers for our time as we delegated our morality to a crime syndicate. Our heritage was supplanted by a committee that called us monkeys and jabbed us with needles. They told our children they were endlessly shrinking inside an ever-expanding explosion where no one could hear you. They sold us stories of endless war while constricting our resources.

Since 1918, a citizen may not possess the feather of an eagle or a raptor. Society required the killing of all raccoons, skunks, opossums, coyotes, and foxes caught or orphaned. We blamed the deer for crossing the road when it was the roads that crossed the forest. We killed our totems mindlessly. We were suffering from wetiko or cannibal psychosis. Witiko were frozen-hearted, man-eating giants of Cree mythology coated in ice. They are the machine that fed on our soul before we awoke.

After the Great Awakening, we discovered the government could run on digital contracts. Ethereum is a blockchain technology that supports our tapestry. With digital contracts, voluntaryism can thrive inside a consent-based government. It replaces the coercive machine with a community-driven project incubator. This is the essence of natural law and government. Bioregionalism is water percolated from the soil. It's a belief that government blooms naturally from the roots up.

America can be a nation of bioregions joined by a people's Bill of Rights. Our nation's militia belongs with the people. Each bioregion can muster their own troops and decide individually on war. Each bioregion can have its own EPA. Each bioregion can have its own Supreme Court. Each bioregion is sovereign land. The District of Columbia has no claim to our consent. Consent is an extension of our will. When we the people reclaim our consent, we walk upright again with the land. The land gives us resources in exchange for our awareness. Delegation is not awareness. Delegation is a form of abandonment.

CHAPTER SIXTEEN
Equality is a Bad Word

Equality is a bad word. It sounds weird to consider but words can be judged good or bad by a simple mathematical formula. Prana is life-force. Does a word extract or install prana after hearing it? How does it affect your prana to hear economic equality? Racial equality? Gender equality? A deficit of equality enlists injustice while a profit in equality enlists guilt. Equality is a machine that harvests sympathy for power. This is a machine of self-evacuation. Once the self is gone, the victim is willing to join the hive. It won't matter what color or side, they just don't want to feel alone anymore. This is the essence of communism.

Equality was invented by psychopaths who pretend everyone needs their help. They insist help only comes from a centralized power that becomes generous and kind after it wins control. Equality insists it will pay after it controls every living piece. Equality insists this is the only solution. Equality creates the need for a solution. Equality is a trap for gullible people who think blind empathy is kindness. Blind empathy is what got us here. Equality is defined in three words, "Eat the rich."

An old friend from childhood spends his days fighting oppression on a white collar college campus under the corporate flag of Black Lives Matter. As a white man with guilt, John is empowered by the solace of racial inequality. I knew him as a boy. To watch him now fully invested reminds me of the true power of mind control. He is convinced equality is the answer. He calls me an ignorant Nazi because I don't agree. He calls anyone who doesn't agree a fascist.

The Technology of Belief

One-hundred-million dead from communism isn't enough for him. He needs more bodies to reconsider. He'd rather punch Nazis in the wrestling ring before looking to the promoters. Injustice gave John a fetish for anger. He becomes trapped in the dialectic. My friend hates my telling of the truth. He calls "mind control" bullshit and thinks himself immune from groupthink. Last week he was shouting a corporate motto from MSM without knowing it. Narratives are designed for the brains who think themselves too clever. Mind control supersedes intelligence. Mind control is the constriction and release of dopamine.

Equality does not exist in nature. It came with the invention of the factory. Man is pulverized and extruded into a shrink-wrapped package and stacked on a pallet. Most of the history of Bolshevism was left out of schoolbooks in high school. Most of the killings and death brought on by equality are ignored in college. The French Revolution failed because of its insistence on equality. The American revolution succeeded because of its insistence on liberty. Colleges won't tell this truth anymore. Colleges say Mark Twain is racist. Colleges want equality, not discernment. Colleges want every book on the shelf to have exactly two-hundred pages.

Equality calls the penis an advantage over the vagina. It insists every white is here to suppress every black. Equality is a narrative of the narcissist to set up a victim paradigm. Equality is cannibalism and we will chew off each other's feet insisting it makes us feel better.

Individuals can be victims but never a group. Being hurt requires a body to experience pain. Any identity movement claiming victimhood is a figment of the imagination. It dismisses the human in favor of the hive.

Most minds won't consider ideas outside of their programming. This is the purpose of programming. It keeps us in our lane. Prana constriction is vital to mind control. When we get sick it weakens the life-force and we feel threatened. When we are self-empowered it grows our energy and we become more compassionate.

Pain makes us selfish. Abundance makes us thoughtful. Our government uses mind control to suppress our prana. This makes us malleable to bad ideas like equality, centralized power, and communism. If John truly believed there was injustice in the world he would pick a victim and fight. Instead he defends a color or a genital. But equality is a religious cult of oppression. It requires oppression to survive. Without it, there is no glue for the congregation. My old friend is a shepherd pretending to be a revolutionary. He never once considered equality could be a bad thing.

CHAPTER SEVENTEEN
The Snake Oil Messiah

Zeitgeist is the mind of God. We insult each other when we assume humanity's thoughts are separate from him. God is the ant dragging a crouton as much as he is the oil tycoon crowned by a secret society. Strange is the lack of information on the richest family in American history. When tracing the source of John D Rockefeller, Sr's fortune, all archives "relating to both the family and individual members' net worth are closed to researchers." That's the story we accept. Truth insists we see the shadows. The Rockefeller bloodline is steered from behind a forbidden curtain. There are meta-people living above the law of countries or birth certificates. These entities employ legal wardens like the Rockefellers to run their farm. Four generations of Rockefeller demonstrated a brutal loyalty to the constriction of energy, mind, and medicine for global power. They were the pioneers of globalism and they never worked alone.

We know a bit about John Sr's father, William "Devil Bill" Rockefeller. He sold potions under the alias of Dr. Levingston. William was many personalities. He was a married man with eight children and a practicing sexual predator. His housekeeper Nancy Brown gave him two illegitimate daughters. The following Spring, when John was two years old, he would watch his father rape another housemaid. Ms. Vanderbeak was in John's bedroom changing linens when William entered the room. He stuck a gun in her mouth. He squeezed her neck like a warm goose and finished her over on the bed rail. It was a May Day afternoon for John playing with his blocks in the hall. John was accustomed to violence. His bloodline had been stripped of morality a long time ago. John would battle depression his whole life. Man is haunted by the sins of his father. We make deals with our children's destiny every moment of our lives. Trauma was a dangling puppet string for the Rockefellers. A secret society with longterm plans turned this into a commodity.

John went to college in Ohio. In his twenties, he started a supply business two years prior to the Civil War. Instead of fighting for the Union, he was poised to sell them groceries and oil. In his very first year, Rockefeller turned a $4,000 investment into $450,000. By fate, the Federal government began subsidizing the price of oil as much as 3000%. All of the profits flowed directly into Rockefeller's pockets. By 1863, the talons of the Standard Oil Company, America's first monopoly, had clenched 90% of the market. In 1911, it was ceremonially split by anti-trust laws forming the seven sisters of oil we know today. This government-enforced break-up was instrumental in making John D Rockefeller, Sr. the richest, most powerful man in America.

April 20th, 1914. In Ludlow, Colorado a thousand coal miners stopped coming to work for Mr. Rockefeller. They had been forced out of their homes and living in a makeshift camp while on strike. Sixty-six miners, wives, and children were gunned down and burned alive on the camp's property. The guards and the state militia were hired and supplied by Rockefeller. No one, including Rockefeller, stood trial for the mass murder.

Deities are ranked by their omniscience, omnipresence, and immortality. The Rockefeller bloodline was omniscient and invincible from the law. Despite the corruption, every generation of Rockefeller was anointed by an eye of providence. Every Rockefeller lived a long and healthy life working for globalization.

Consider this. Wikipedia politely calls globalism "internationalism." From Kissinger to Rockefeller to Soros – every proponent of globalism has been a rich white guy. No one calls these people White Internationalists.

"Globalists are White Internationalists." - Yours truly

Eleven years prior to the Ludlow mine murders, the richest man in America became the godfather of compulsory education. In 1903,

The Technology of Belief

John D. Rockefeller spent $180 million dollars building a factory for the American mind. The United States Congress adopted his plans for the General Education Board with a mission statement, "the people yield themselves with perfect docility to our molding hands." Compulsory education established the government as the shepherd of the mind, and the people as its flock.

"We have not to raise up from among them authors, editors, poets or men of letters. We shall not search for embryo great artists, painters, musicians nor lawyers, doctors, preachers, politicians, statesmen, of whom we have an ample supply...The task we set before ourselves is very simple as well as a very beautiful one, to train these people as we find them to a perfectly ideal life just where they are."

All the propaganda we've swallowed comes from centralized compulsive education and media. History is written by a single winner and we are trained to repeat their jingles. The richest man in the world wanted something beyond money. He wanted obedience. In 1923, nine years after the coal miner massacre, his education foundation would go international. His addiction to control superseded human life. Rockefeller was turned by the same gears that turn modern CEOs like Jeff Bezos. Men at the top are driven, broken, and predictable. They have been bred with a fetish for asphyxiation which makes them an ideal asset.

The boa constrictor does not suffocate its prey. It renders the victim unconscious by eliminating its blood flow. This tactic made Rockefeller a success in oil. It made him a success in medicine when he declared an economic war on plants. The 1910 Flexner Report set nationwide protocols for mainstream science in teaching and research. As a result, nearly half of all medical practitioners were jailed or closed down. Rockefeller created the American League of Municipalities and the American Association of State Governments to directly influence small towns and state legislators. Under his nationwide influence, all doctors and hospitals would require licensing through a Rockefeller controlled board. This affected all hospitals and teaching institutions nationwide. Combining these two influential organizations with his General Education Board reveals Rockefeller's desire to monopolize

America's energy, education, and medicine.

The term conspiracy was applied by the CIA to cover up the truth of JFK. The term "quackery" was applied by Carnegie and Rockefeller to centralize allopathic medicine. Allopathy is the treatment of the symptom. Homeopathy is the treatment of the cause. Treating symptoms is an industry. Treating the cause would kill the treatment industry. The science of radiation and pharmaceuticals is an economy. Probably more than anywhere else, mind control is rampant in medicine. One is dismissed with ad hominem the moment a thought crosses the line. Writers like E. C. Mullins who exposed the 1910 Rockefeller takeover of big medicine in his book Murder by Injection are blackballed in public with labels like "propagandist" and "antisemite." This is the barbed wire fencing of mind control. It's as easy as an emotional reaction from being shunned or praised. Propaganda is the product of an allopathic society. Dissonance is sold as a cure for dissonance.

The Civil War was really the second Revolutionary War but the people lost. That's when the instrument of the Rockefeller family emerged. The hidden strings aren't so hidden if you know what to look for. American freedom is the Gulliver who enjoys life on his back. He is subdued by the technology of the smoothie. Where slavery becomes palatable the whole world falls.

Why would God work through Rockefeller's constriction? Where is justice in a world of energy slavery and profiteer medicine? The answer lies beyond the ball fields of good and evil. In the parking lot, our consent waits in the car while it's raining. Man is brought to this game to discover a line separating trust and slavery. If God was recruiting the perfect team, he'd chose the players least susceptible to corruption. We betray ourselves giving our consent to a stranger called government. Losing the sovereignty of a state's airspace means losing the sovereignty of our minds. Our history has been fully corrupted by their propaganda. The people who invented globalism burned Atlanta and labeled the south a tribe of traitors and racists for complaining. The mind control will continue until sovereignty improves.

CHAPTER EIGHTEEN
America Believes

Circumcision is sanitation. One white guy discovered America. Bread is a food group. Humans are fornicated pond scum. Pizza is a vegetable. We called the moon from the White House. Oil comes from dead dinosaurs. Inanimate metallic objects commit murder. Foreign homicide is patriotic. Vaccines are safe and effective. Abortion is health care. A two-party system is representative government. Dandelions are weeds. There is no such thing as inflation. The USS Liberty was an accident. Income tax pays for our roads. Plant medicine is a felony. Collecting rainwater is a misdemeanor. Everything is racist. Magic bullets turn in mid-air. Building 7 was tired. $2.3 Trillion was missing. Iraq had WMDs. Genitals are a thought experiment. There is nothing to see in Antartica. Gravity is magic. Aliens come from outer space. Geoengineering will save us. 5G is harmless. 6G is better. Corporate news is real.

CHAPTER NINETEEN
Secretions of the Spider

Fashion is the pinnacle of mind control. War is reduced to a lethal wardrobe decision. If we fought for self-preservation, no soldier would breach their own border. If we fought for ideology, we would quiz each other across the trenches before shooting. In war, it is exclusively the fashion choice of the enemy that makes us pull the trigger. Fashion is homicidal. It is the ultimate mind control. Men in orange jumpsuits on the side of the road avoid our pity. Metallic stars on someone's chest avoid our scrutiny. Fashion is the fabric of propaganda. Fashion is human possession in a costume. In it, man transcends his identity and loses the weight of his ethics. It grants him the right to call war hell instead of murder. Costumes make us amoral in politics, medicine, and justice. We are a society of blue and white collars arguing for red and blue neckties under the authority of white and black robes. We wrap our necks and nipples in silk. We slide platforms under our heals to harness the power of our costume. We withdraw life-force from monograms and sigils. Fashion makes us feel less hairy and more exotic. Fashion is the secretion of style and style is a secretion of belief.

International artist Louise Bourgeois was more than a trendsetter. Heralded a champion for establishing a home for wayward girls, her sculpture work captured the beauty of female suffering. Born on Christmas day in 1911, she grew into an artist whose work focused on scenes of confinement, mutilation, and torture. She expressed the truth as a witness. She cherished an appetite for horror and the elite fed it. Debuting in 2001 over Rockefeller Plaza, her twenty-foot-tall

The Technology of Belief

bronze spider Maman was one of the largest ever erected. Bourgeois was the Queen of Spiders and darling of the art world. Corporate grants across the country consented to bring her deep black menacing creatures to overpower the landscape. This was a flaunting of elite power through the world of art. We are so deep in the trance of safety we would never dare to see the truth. Human slavery runs the world but the hairy belly of this spider is too big for us to face. We are the lightning bug blinking inside a silk cocoon.

Trends are measured by their ability to captivate. Like a spider's web, trends are dangling silk from the artist's charisma. Trends glimmer in the light and gobble in the dark. They capture our identity and wrap their sticky legs around our wings. We become a slave to style. But the sticky silk is too tempting to admit we could stop. Style drains self-worth with the venom of heroin. We are mesmerized in the comfort of pre-charged prana. We are enriched by what we know makes us a slave.

You can buy a high-quality newborn on the black market for $24,000. Most of the babies bought are used for ritual sacrifice. The ones who don't die suffer something blacker. How does someone become a John or Tony Podesta? How can two brothers gain so much influence while being addicted to human flesh? Cannibalism is an anti-baptism. Instead of being reborn, one commits living suicide. This first supper is the final evacuation of boundaries and morality. In Tony Podesta's home, the sculpture of a body decapitated has been dipped in gold. It shares the pose of the cannibal Jeffrey Dahmer's victim. Bourgeois made this sculpture and several more like it in his honor.

Of her work, *Destruction of the Father*. Bourgeois reported:

"*A kind of resentment grows and one day my brother and I decided, 'the time has come!' We grabbed him, laid him on the table and with our knives dissected him. We took him apart and dismembered him, we cut off his penis. And he became food. We ate him up... he was liquidated the same way he liquidated the children.*"

Foster care is an organic market for child trafficking. The industry needs children to survive. Rachel was an orphan and a documented flight risk. This made it easy to disappear from the system. She entered the widow's cage at age twelve. She was recruiting for the program. Bourgeois had been hatching spiders for years. Rachel would graduate from her program as a helper. Outside her cage was a long table with benches. In the pen next to her was a butchering chair. Rachel watched what they did to people in the chair. Then she watched them eat their work at the table. They pulled out the entrails to stretch across the room while the victim was still conscious. Living bodies hung were from hooks and swung from the weight of their own breath. The first few days Rachel feared she would die. Then she caught herself begging them to end it. She taunted and cursed but no one responded. Day after day, she was invisible. The suffering eyes of the dying were the only attention she received. Victims in the web talk to each other in their glance. They dare not talk. The vibrations ring the queen. Tuning a slave is the art of secretion. Louise's cage would recalibrate Rachel's appetite just as it did to the brothers. A spider knows her business.

Our schoolbooks insist America ended slavery during the Civil War. But real slavery has nothing to do with chains. Slavery is the hijacking of identity. Slavery is controlled trauma. Slavery replaces identity with an addiction to security. A slave is predictable. Their labor is harnessed without the need of whips or chains. Like dogs, men are made slaves. Once you control their wiring, you can trust them with billions.

Epstein woke up again with his fingers clenching an erection. Someone was buzzing at the gate. He walked to the kitchen with his boner leading the way and checked the camera. He was disappointed by only seeing three of them at the gate. He was promised four or five. He was hoping for eight. It wasn't about the money; Epstein had millions. He needed the fresh meat. Epstein was a serial defiler. Once he tainted a girl his own disgust would turn him off. He was a perfectly broken motor and a world-class cannibal pimp. He pulled $1200 out of his safe and separated it into three piles. $200 for each girl and $800 for the commission. He posed each of them in his room on a white rug and masturbated as they watched him grunt. He enjoyed the exchange of energy as each girl clung to the remnants of their dignity. He sucked the prana right from their eyes. Dignity is a holy oil and Epstein loves to drink.

How does a Math teacher with no college go to managing Rockefeller's money portfolio, a seat on the Council on Foreign Relations, and a member of the Trilateral Commission? The man who put Jeffery Epstein in power was Robert Maxwell, a Mossad agent. Publisher and member of UK Parliament, Maxwell "drowned" in 1991 before it came out that his pension group had been robbed of millions. Maxwell was behind Mirror Group Newspapers. Think mirror. Epstein's mentor was Steven Hoffenberg. Like Maxwell, Hoffenberg made millions from investment fraud. The man who made Epstein and the man who taught Epstein both owned big media. Both were criminals. The power to expose is the power to extract. Someone invested a billion dollars into propping up Epstein. They considered him a reliable motor. He was recruiting underage sex workers and using them to service world leaders like Bill Clinton and Prince Andrew. Despite his billions, Epstein was one of many low-level spiders. Epstein is a product of Mossad. Mossad is the CIA of Israel. To us its an NGO living inside the American government. Either way it's a tapeworm we pay to chew our legs off.

Epstein is to Bill Clinton as Kissinger is to America. Obama, McCain, Bush, and Clinton were assets from a closet full of Manchurians. Each of them was dressed by the same butler and paired with appropriate shoes and tie. The uncomfortable part is realizing there is more than one spider. Kissinger is hairy but even at his size, we know, he is one of many. Epstein's boss was Les Wexner, a billionaire Jewish mobster mogul behind brands like The Limited, Abercrombie & Fitch, and Victoria's Secret. Wexner runs an NGO linking Israeli politicians to Harvard and by extension the Brotherhood.

Do you get it yet? Do you understand human slavery goes all the way to the top? Don't even try and swallow this pill. It won't go down. It's too much for one person to digest. Just take a nibble and understand this will never be accepted by the mainstream. Like the land of the Aztec, we live in a culture of human sacrifice. The Clinton body count is a huge list but it only includes deaths on domestic soil. Despite this, 56 million people were certain she was their gal. This is nationwide, centuries-old Stockholm syndrome. Denial is the widow's cage where we cheer for anyone who gives us attention.

The elite are mangled genetically so the machine can be installed. Most have been bred for centuries to be brutal through cannibalism and torture. It's why NXIVM was running eleven elite preschools around the world. Most elites leave their own families by age four. Boarding school is professional trauma disassociation. The answer to what makes so many people loyal to depravity is trauma. Trauma is an epigenetic pyramid scheme. "Where is the injury if an adult has sex with a child and [the child] enjoys it?" the NXIVM leader wrote. "What is the difference between the child being tickled and being stimulated." This is how relativity works. In the land of lies, children become sex toys. The NXIVM scandal goes all the way up. There are 15,000 members of NXIVM worldwide. Raniere knew what made desperate people tick. The ranchers running the farm know this trick, too.

Remember the blueprint of mind control.
  1. Break target.
  2. Constrict aid.
  3. Harvest life-force.

Dopamine supersedes intelligence. Mind control is its secretion. Satanic ritual abuse merges the feeling of "I miss you" with "You hurt me." These two feelings are combined in the same chemical feed and complete each other. As Pavlov explained, the chemicals prompt each other by their anticipation. This is why survivors don't heal so much as grow. Trauma is the art of pruning the bonsai. It's the same on any livestock farm. Human slavery is a product. The market holds a monopoly on the world through infiltration.

We are tuned from our youth to develop a certain chemical palate. Mind control has nothing to do with thoughts and everything to do with appetite. Rachel will join the brothers for her first meal eventually. They are waiting for her to beg. Her words will be her inversion. She casts herself into the pit. Her brain is rewiring every day in the cage. The lack of food puts her deeper into the trance. Rachel will grow to love catching flies.

Cannibalism is the crossing of a border. The sympathetic and parasympathetic systems have inverted. Cannibalism isn't eating just as rape isn't sex. These acts are both a self-evisceration. Cannibalism is a devotion to the anti-self. It turns brothers into addicts, psychotics into international artists, and puppets into Presidents. There is power in the abandonment of boundaries. The only slavery that exists is our addiction.

Trauma and power merged in blood thousands of years ago. Evil has its roots deep in man's soil. Evil is the art of evacuation. The soul is slowly secreted. We do a disservice to pretend evil is a living thing. Evil is a lifeless ever-emptying hole. There's nothing living about it.

CHAPTER TWENTY
Corporate Pride Month

I saw a yellow jacket today through childhood eyes. I was precious in the trance. We are fully immersed in our senses. We tune so deeply into a signal we disappear into its verb. A child is lost in the department store as he ducks inside a rack of raincoats and jeans. His mother is calling only a few feet away but the boy is 20,000 leagues under the sea. He is hunting the Kraken. We are older now. But life is still lived through a trance. Adults are jaguars in raincoats hunting in JCPenney.

Man is a sensual creature, fully immersed in the nervous system. The deeper one is seated, the stronger the charge from his soul. Sex is more than sensual. Sex is alchemical. Sex opens the stargate for life. Sex magic is real whether you believe in magic or not. The sperm and egg are a circuit in the electric universe. A corporate LGBT program is stifling this magic in the height of summer when minds are out of school. There are socio-political reasons to push and install corporate Pride Month. These reasons are eugenics and victimhood engineering.

You can be gay and object to Pride Month. You can keep sex intimate and special between consenting adults or you can use it as a tool to mock and provoke people while labeling them bigot for not cheering. If we want to let our pride out let's be real about it. Corporations are pushing sodomy and scissoring on the average of every twelve day of the year. This makes an impact on the masses. We are called bigots for pointing this out. I have been told it's a display of pride to say the word gay and a display of hate to say the word sodomy. Why? The definition of sodomy is sexual intercourse involving anal or oral copulation. If there is shame around that word, Pride Month should be working to dissolve it.

Could being gay be cultural? Why is it not acceptable to ask that question in public? We are rebuked for asking what determines sexuality. We are told no one would ever choose to be gay. It is explained that persecution is so terrible no one would freely choose that path. Ask the politician how weaponized persecution is a motor. Study Stockholm syndrome or anything counter-culture and you find solace in the underground. We are told gay must be genetic in an effort to take away our volition. We are told we can't help what we like and there is nothing we can do about it.

But the truth isn't stupid enough to bite. What we watch and ingest will always come out in our habits and proclivities. The trance we inhabit now is determined by all of the trances we lived in our past. Media sets our pace like a psychic metronome.

In today's world, the word homophobe means someone who doesn't believe in victimhood. Calling someone homophobic for rejecting corporate Pride Month is like calling someone white-phobic for rejecting Nazis. These phobic references are mental land mines and the people that plant them for you to step on are the ones who profit from core shame. Shame is a commodity to the immune and psychopathy is a human antibiotic.

We are responsible for what we do as zombies. If you grew up victimized by the system or through epigenetic trauma you will seek shelter inside a group that caters to victims. We place victimhood on a pedestal precisely so we can grow more of them. Insanity is a crop harvested every April 15th. We just call it freedom. Victimhood is the pit you'll never escape because compassionate people keep pushing you back in. This is the horror of blindfolding the empathy. We distribute victimhood on a tray like free samples at the food court.

Gay is not an identity. People who lack identity don't know the weight of their own name. Mind control is an instrument of shame and fame. Shame is applied to anyone heterosexual while fame is applied to anyone homosexual. These are both forms of abuse. This is the blueprint of triangulation.

Legislating hate laws declares a hierarchy of citizenry. The rainbow flag should not be a seating chart. This hierarchy will always breed injustice. Anger is a natural reaction to injustice. But anger breeds victimhood when we call it toxic. There is no LGBT community. There is a universal pursuit of happiness shared by all people. No one objects to that.

If America was homophobic we'd throw people off buildings. Instead, we dedicate a month to its pride. This isn't equality nor is it justice. It's called gaslighting. There are places in the world that are homophobic. They throw people off buildings for being homosexual. We don't criticize them. We can't vote our way out of this mass psychosis. People should be fighting psychopathy not homophobia. Homophobia, like racism, is corporate mind control.

Pride Month is a cover for psychopaths to do what they always do. Push infanticide, pedophilia, disassociation, and eugenics. If this was about pride, we'd be ending genital mutilations and stop calling them circumcisions. Pride Month is a form of empathy-jacking. This is a technology of belief.

Corporate Pride Month shows us the mechanics of controlling a populace. It's no different than a cowboy managing his cattle. Pride Month is the human version of branding and gelding. It marks each of us with a different sigil and corrals us on an island of flies. We are the lords of trauma. It's leathered our flesh so thick we can't remember our skin. Anarchy is too heavy a yolk for our sovereignty to bear. We want to rest here a bit longer pecking for worms on the animal farm. May the protein give us the motivation to finally rise.

CHAPTER TWENTY-ONE
Sins of the Father

The opposite of epigenetic trauma is wisdom. We inherit gold through our genes as often as we inherit debt. An unconscious life is genetic usury of your grandchildren. This is the birth of vampirism. But wisdom is gold when it comes from the skin. Charisma is the vibration of DNA. These threads connect us to ancestral lines. The sins of the father are felt through four generations as the mitochondrial DNA are shed three times in a row. It's a trinity baptism. Like all of our rituals, we cast our spells in threes. In the name of the Father, the Son, and the Holy Ghost. Life is the inheritance of wisdom and trauma. If reincarnation is real, surely karma is recorded in the DNA.

Recessive genes appear to skip a generation by not showing themselves on the surface. They are dolphins diving and leaping above and below the waves. We carry flags from our mother and father, but the male Y chromosome is inherited paternally. In this way, we shed our past like a dead snake losing its skin on the thorn. Where you tend a rose, a thistle cannot grow.

Shannon was the size of a munchkin. Her eyes were deeper than the back of her head and her hair was as white as a storm. Ruled by Saturn, she glowed purple in the dark. She was twisted perfectly by the sins of her father. He was jealous of her survival charade. Her whole life had been marinated in suffocation. She saw the life of a rabbit and decided to bolt. She thought she could outrun the endocrine system. Shannon showed me a world I hoped was fake. She sat on the floor of my apartment when a burst of light erupted from a speaker that wasn't plugged in. She had been staring at it for some time. She didn't flinch when it burst. She looked up at me to see if I would blame her. The walls in that place were crumbling with purpose. I clawed the plaster to get at its brick. I painted my first mural on those walls in the bedroom that turned into a kitchen. Sex for her was more like feeding.

Shannon was trapped in the trees surrounded by marauding disbelievers. They taxed her sanity with a thousand cuts from a thousand pairs of scissors. She was a gypsy who grew up on a pedestal of tacks. Don't ever turn your back on a ritual survivor. If you gift them trust they lose respect for you immediately. And so they should. You have no idea what they are capable of thinking.

I would meet Shannon twenty years later in the town of ashes. She burns there now like a candle casting its spell on the dark. She keeps a boy on a rope like a pocket watch that needs winding. She barely notices the time. Before he came, I dipped into her ladle and drank her gnosis. The sweetest waters collect in the cups of her collarbone. I sip that juice like a hungry hummingbird. We poured each other tea and spoke of psychopathy. I wondered what makes a heart beat after the chest is evacuated. The answer is the tremors of sadness from lungs who lost purpose. I track her brow as it furrows. She is more than a catch, she is the ocean. She is a pirate pretending to be a surgeon in a room that's too messy to clean. She could pop a cork with her teeth if she drank. She had strict rules. Every drug is a monogamous relationship.

We demand healthcare from people who tell us plant medicine is a felony.

Cannabis is a jealous lover who wraps her fingers around the skull. She is a gentle giant showing you where to look. Dopamine comes from the retina. What we see is a chemical illusion of what we want to comprehend. We are deciphering chemicals from each other's brains with bent fishhooks. I lost her from seeing too much. I knew her like a boy knows the cave behind the ballfields. Shannon could never come home. Exposure would burn her at the stake. This is the safety of turtles and lizards. We develop a fetish for scales and armor. We find secrets to build our nest where nobody looks. The most powerful among us are the hardest to heal. They carry on the longest without limping.

Shannon is the sin of her father. Her father is the sin of another. We are trembling reeds of what happened to someone a long time ago. We are vacuum tubes and milk bottles left on the porch step. We are the ice cream truck combing a neighborhood like a hungry lawnmower. If I missed Shannon now, it would hurt. If she missed me, she would crack open. Secrets keep us invested in a committee of one. Self-loyalty is more important than society says it should be. Lies are a lonely secret. If the secret brings you closer to yourself you should keep it.

CHAPTER TWENTY-TWO
Flat Earth Karate

Atheism is the cancer of thought. The Big Bang was the invention of Atheism. It's the runaway brain devoid of deep breathing and guts. Atheism is a swollen head in a jar. Faith is the function of intestine and lung. Our lungs collapse on the faith they may return. Guts shoot for the three-point shot from half-court as the final seconds dissolve. As the bumpy orange ball spins through space hurled at a metal ring, the receptors in the brains of both teams hang on the outcome. Half of the players will be rewarded with dopamine; the other half will be placed on restriction. The memory of this game will forever yield a chemical bump. Mind control is the cessation of lollipops.

    We are all suckers. We've been rewarded for regurgitation from the time we were young. In a classroom, with a globe, we repeat the fantasy as fast as we are told. The pop quiz is Pavlovian. Our mouth waters at the rank of letters augmented with a plus or minus. F is failure. A is applause. We gain victory in school, not wisdom. That's why every question on a test has one of four answers. Like rats choosing water over salt, the natural mind will eventually find the truth. Flat earth is a pose of seeking in the dojo. I am not here to tell you the shape of your world. I'm here to show you how to look.

    What happens to the spirit when it hangs on rumor? NASA's words dangle us from cliffs as asteroids careen through space to kill us. Flat earth is a stance of protection. It's a technique to attack dissonance and lies. Flat earthers know most of what they've been taught by compulsory education is a lie. Propaganda is a trance

The Technology of Belief

installed by school bells and rectangular pizza.

A good defense is never a distraction. Flat earth cuts through lies like a plasma cutter. It adjusts your stance in the dojo when you're tired of getting punched by lies. Globe earth is globalism, special relativity, godless vacuum, and NASA. All globe earth has in its corner is ridicule from the majority. The essence of scientism is a doctorate in ad hominem.

"It is the mark of an educated mind to entertain an idea without endorsing it." – Aristotle

Flat earth is an antibiotic. It creeps into your system and obliterates propaganda. Good or bad, you are left with nothing to lean on except direct experience. This is a powerful tonic with the tang of castor oil. Flat earth is difficult. You are rebuilding your world from scratch, which requires calories to keep it alive. You are surrounded by people who want you to stop. Your stance makes their boot's quake. That's what happens when you have no foundation.

Practice flat earth like a posture. Fill your cup from the inside. No one can tell you the shape of the earth. Globalism is a powerful force. It requires something even more powerful to defeat it. That power is the engine core of your identity. Globalism is the lie pushing against the only real thing you know. Your experience is who you are. Belief is where you place your trust. Belief is a precious energy. Globalism teaches you to doubt the senses. Globalism teaches the self can't be trusted. Globalism is the gelding of the cord connecting you to the earth.

It's impossible to believe in evolution when you stand on a flat earth. Your feet are planted on the ground of an electric universe. You are the completion of a circuit. You are the divine spark. Flat earth places you back in your system. This is your center of awareness. We approach the world from an assemblage point. This is our needle of awareness. Globalism places this high above the aluminum clouds outside the spinning earth. Under globalism, there is no such thing as direction. With globalism, we are the dead dinosaurs of the future and the virus of Gaia. Globalism is disassociation. It's the turning of self from truth. Globalists fight truth with ridicule. Globalism enslaves with relativity and materialism. Globalism is a welcome to the machine.

Slavery is training wheels for the sovereignty of thought. As man evolves, he sheds the desire to be controlled by a master. Slavery comes naturally to all of us. It's how we learn to stand and walk. The purpose of life is the transmutation of slavery to sovereignty. This is why we are born co-dependent. Some flowers don't open. Many prefer slavery as a kind of childhood for grown-ups.

Globalism is the grid; flat earth is the disconnection. Rebuilding your cosmology directly from your experience. Show creation you take this puzzle seriously. God doesn't need you to be right. God wants you to try hard. Use your senses to discern a world free from propaganda.

No one will be rewarding you for flat earth. The ups and downs add up to a net loss of energy calories. We are trapped on insanity island waiting for the plane to tell the boss. Flat earth changes you internally. It builds new levels in your underground fortress. The soft belly turns to rock from the arrows and observation. The gut grows an armor thicker than any thief who tries to steal it. Flat earth is you, alone on a mountain, peering through the curve that used to be there.

CHAPTER TWENTY-THREE
Trumps Flów State

Flow state is when you stick your hand out a moving car window and let it swim in the airstream like a dolphin. You are in effortless creation with the world. You're no longer thinking; you are experiencing. Your identity as a noun is transformed into a verb. Even you don't know what will come out of your mouth next. You have no fear of the freedom of flow. Flow state is the optimal mode of trance. In a flow state, you tap into the environment and find abundance. In a flow state, awareness is a surfboard.

Everything we do is a trance. Alertness is a trance. Joy is a trance. Destiny is a trance. Apathy, fear, guilt, and self-hatred are trance. Some trances are good for you. Some drain you like a battery. All of our energy goes somewhere. We are corralled by the fear of our choices unleashing a plague of locusts. In our hearts, we check each of our purchases at the exit doors of a buyer's club. We've been taught to treat our own vibration like a shoplifter as all of our profits are lost to control. This breaks flow state. Mistrust is a hefty tax to the life-force. Everyone is afraid of being the person who sucks at karaoke without knowing. But that person has an abundance of flow. Their ignorance makes them invincible. Trump has this flow state. It's nine o'clock on a Saturday and Trump is a nasal-ridden goose singing Billy Joel. And in his mind he sounds perfect.

We forget politicians don't always speak to their constituents. Trump's success is letting him speak directly to history. Trump wants a legacy that's bigger than all the elephants. His ego is the best thing that could have happened to America. People hate the invincible for being invincible.

Back in 2015, onstage at the annual CPAC conference, Trump said something very interesting about Bill Clinton. "Nice guy," he said. "Got a lot of problems coming up, in my opinion, with the famous island with Jeffrey Epstein. Lot of problems."

Epstein's boss, Les Wexner, was a fashion mogul for Victoria's Secret, Limited Brands, and Abercrombie & Fitch. Wexner's foundation created a bridge between Israeli politicians and Harvard. The truth is a wet snake under a loosened rock. Epstein, Wexner, and Seth Klarman, all drip in the slime of Israeli deep state. These men operate through charities like Ohio State's Columbus Foundation, OSU, and Gratitude America, Ltd. They have lesser children working in investment fraud and mass media. Soldiers like Robert Maxwell and Steven Hoffenberg are Epstein-funded media operatives and investment fraudsters. Epstein wasn't an investor as much as a slave trader. He's been traced back to the Imran Awan and Dynacorp as early as 2007. These links make Epstein's case potential good news for an America controlled by blackmail and addiction. Why was Trump willing to lose $9 Billion in support to tell America our elections can be rigged? Why would two Republican Jewish billionaires turn on Trump after he gave them Jerusalem? This is what the Israeli Deep State looks like. This is how deep their pockets go.

It's impossible to criticize Mossad in public. People are simply too brainwashed by the myth that supporting narcissism and genocide gets a golden ticket to heaven. To my critics who say I'm anti-Trump, anti-Christian, anti-Q, anti-jew, or anti-American – you're peeing in binary pajamas. I have always been loyal to anti-slavery. I have always been anti-corporation. The District of Columbia is the opposite of both. Every sentence I write continues to bring us back to sovereignty. This is something we lost in the not so Civil War.

The Epstein case won't solve government. But it can solve ignorance if we push hard enough. Solving ignorance takes a strong ego. This is why society keeps the ego under quarantine. When you have a strong ego like Trump, people don't have to like you. Your behavior is no longer motivated by external forces. Freud insisted every man wanted to have sex with their mother. He shamed his ego

The Technology of Belief

on paper and academia praised him in their books. This is why we flog ourselves in public. Shaming the ego makes us controllable.

Trump is an alchemist, not a politician. Picture how expensive it's been to hate him. Picture the size of the container of spent life-force from the Love Trumps Hate campaign till now. Every day, haters spit spent tobacco into a jug. Trump hate is painfully slow dehydration. Every day the jug gets heavier from the load. That weight was life-force a long time ago. Now it's a tax paid through lips forced to spit the name, "Donald Trump." From love trumping hate to loving Trump's state, ego and supremacy can save America.

Many find the idea ridiculous that bad guys are going to put each other in jail. Let me make Trump's art of the deal as clear as crystal. None of this is going to end the cabal. At best, it will simply change hands for a few generations. Our world is run by a tapestry of international gangsters. We lie to ourselves by pretending things will be different. Voting is a symptom of denial. So too is hope. What makes Trump different is a strong sober ego unruled by a desire for prepubescent children. That's what sets him apart. At least we hope so. This has always been why the media is against him. He literally scares the shit of pedophiles.

Trump has traded America First for Greater Israel. Trump gave his daughter Ivanka to Netanyahu's protege Kushner as a blood contract. This marriage stops another presidential ritual sacrifice like JFK. Ivanka's ovaries stop the second magic bullet. Without her, that contract would have been open the moment Trump won the election. When you think about it, Ivanka Trump has been the most important piece of this administration. She is an elite living under restricted autonomy. Her oars are too deep in destiny's water to steer independently. Ivanka is moved by the currents of a bloodline that launched her ship a long time ago. She's been in training since she was four-years-old. Her children and grandchildren have already been moored to their future. Thoroughbred humans are too expensive to run free in the rain. Trump can speak directly to history thanks to her sacrifice.

Listen to Trump's snake speech again as if he's talking about Israel and your ears are listening from sixty years in the future. We are ruled by our genes and the bloodline is a crucial Achilles' heal. Marriage has always ended the strife of kingdoms. Marriage has always been the perfect art of the deal. We are ruled by the elite because we are too afraid to rule ourselves. We entrust each other under a bloody umbrella. Merging the tarps make it too expensive for any camp to betray the other.

Trump is competing with his own image now. He wants to cast a long shadow over history. Ego is a predictable white knight burning in the dark. Ron Paul was too principled to pull this off. Ross Perot was too moral. Ego is the only thing that could kill the dragon of Israeli deep state. Understand that this administration works for Greater Israel now. Understand it's all been a part of the deal. This is what Q means by "saving Israel for last." No one is sneaking up on Mossad. Trump enables Jerusalem to enable America. If he doesn't, he goes down in history as another footstool. Here's to Trump's ego feeling too important to risk that kind of tarnish. Perhaps ego, self-supremacy, and ovaries can make America great again.

CHAPTER TWENTY-FOUR
Billion Dollar Liars

When I read anything technical from Epstein's "science" ventures it reads like fluffy bullshit you'd feed to a capital investment group. It's a challenge to take it seriously. I watch the big words slowly sink from a lack of concrete. Ask anyone in venture capital and they will tell you most of their job is propaganda. They won't call it by that name though. They call it cutting-edge innovation to make it sound better.

It's important to remember special relativity has been debunked by quantum mechanics for some time now. We even find things that travel faster than the speed of light. Still, society's movers keep propping up meat puppets with a reverence for Albert Einstein. The superstars of science are no more talented than the superstars of Hollywood. Epstein, like Einstein, was an actor playing a part. He was not a billionaire investor pushing the limits of artificial intelligence and biology. He was a criminal committing international human trafficking, rape, blackmail, and fraud.

Scientism tells you the chance of you being born were 1:400 Trillion. This is propaganda. The chances of you being born were 1:1. Half the people who tell you about cosmology were on Epstein island. If they can't discern morality, why are we trusting them with our galaxy? Elite academia is a fraud run by billion-dollar liars. We have much to unlearn.

In 2002, I wanted $200k to develop a WordPress platform before WordPress had been invented. I had a working model of the program we had been developing for over a year. I knew its weaknesses and strong points. I was ready to take it to the next level. My mentor took me to Raleigh and got me an audience with a capital investment firm. We prepped for three months but never showed any code. Harry explained, "The code will only distract them."

I listened to Harry. He was a successful developer with experience in tech. He was kind enough to open the curtain for me and show me how it worked. We didn't ask for $200k. Harry said we needed to ask for $7 million. I thought he was joking but nope. I had to practice saying the words, "$7 million" in the mirror just so I could get it out with a straight face. We didn't need that money for code. we needed it for the illusion.

Harry didn't understand my code. No one in Raleigh would ever see what it could do. No one asked for the good stuff. No one wanted to come into the lab and have a look. My idea was reduced to a commodity. Whether it worked or not was moot. In this world, ideas are rated with clout based on who seems interested multiplied by their net worth. That was the only thermometer that mattered in venture capital.

Harry understood if we could get a big fish to sniff our crotch everyone would be hooked. A $200,000 idea was not as powerful as a $7 million idea. I can't help but think this is exactly how Epstein worked. His billionaire persona was the only thing he needed to bring to any room. Any substance in his ventures would be a distraction. Study Epstein's character and you see a list of people explain how he never read or immersed himself in work.

The novel *Atlas Shrugged* exposed this phenomenon of intellectual cannibalism. The copper god, Francisco d'Anconia, could move markets by snuffing someone at a party or lingering with someone else too long. Look at how our market dips get blamed on algorithmic reactions. Humans are even worse at this than machines. All Epstein studied every day were pre-pubescent girls and persuasion.

Harry and I didn't get the money. The excuse they gave wasn't even valid but I was denied an audience for my rebuttal. My facts would have only gotten in the way. On the drive back home, some girls in a VW van invited me to join them at a weekend contra-dance. I was driving alone in Harry's BMW. Those girls treated me differently in his car. They trusted me without blinking. They never

The Technology of Belief

asked to see my lab. Their attention was satisfyingly creepy. I had fun dancing. Today, I drive a '98 Honda Odyssey. It's my cloak of invisibility. This world is made of smoke. People like Epstein are burning sulfur.

Epstein was not a brilliant scientist. He was a fraudster. Fraudsters spend all their time perfecting their art. Honest people forget the depravity in lying. It becomes fun for the inverted. It becomes a lifelong sport. To watch people believe your lie is to hold power over them. You know something they don't. This is the energy of elitism and the magnetism of the secret society.

We forget how profitable it is to lie. We forget how rare it is to find integrity. John Galt was a character in Ayn Rand's novel, *Atlas Shrugged*. Galt was the inventor who disappeared. He saw the system for what it was and left it behind so it could die on its own. Galt shrugged. He understood the rulers of the world were already doomed. Ayn Rand is one of the most shamed and dismissed authors in academia now. They call her elementary in an effort to reduce her. But elementary is when we learn every color. Rand knew the color of abandonment. Ayn exposed its intellectual cannibalism. She painted as it dripped its grease on the table. She believed in the power of man's spine. She believed in the salty sweat collected under his nails.

People with spine smell Epstein instantly. The smell of money does nothing to mask the quaff of his taint. These are billion-dollar liars. Our government and corporations are full of them. Epstein was a complete fabrication tapping into the Einstein archetype and riding its coattails all over the world. Jeffrey Epstein isn't a genius. He's a shaped tool trained to drive a psychopathic machine.

Our world is full of amazing people and you are one of them. I don't mind confessing to you my motives. The taller you stand the better I get to enjoy my sovereignty. Your psychic health benefits me tremendously. It benefits all of us. Let's stop pretending the elite are smarter than we are. The elite are simply more willing to lie. That's the difference between us and them.

CHAPTER TWENTY-FIVE
Definition of Evil

Evil is a moat filled with bile. The bile's not black it's brown. It isn't warm, it's cold. Evil is congealed. It's filled with crusty teeth and the tangled hair of children. Evil's drawbridge is lined with spent teddy bears left in the rain on a roadside grave forgotten. Evil is not a reveling red demon. Evil is not a hot sauce. Evil is the opposite of passion. Evil is the cobweb we neglect in the corner. Evil's defensive moat is our false perception of how it looks. Evil is not a burning fire. When we say the word evil we picture a red man between the head and feet of a goat. But goats are God's creatures. And man is God's vessel. And red is the color of living blood. None of these are evil. Evil is an ignorant machine that gobbles tears. It's an abandoned wood chipper running in the playground. It's apathy measured by the cold degrees of delegation. Evil is a chain of command. Evil is a spitting abandonment.

It starts when the body is called dirty.
  It's sanitized with a mutilation.
  The shame makes you itch.
  The injustice makes you agitated.
  The drugs make you spin.
  The music sets a rhythm.
  The actors take your hand.
  The self leaves the body.
  The machine takes the wheel.
  But it won't. Because it can't.

The Technology of Belief

Not when you love you.

Let's define evil simply. It is the eviction of life. It is the hollowing out of one's spiritual core. This includes psychopathy, sociopathy, and narcissism. It's the soul that's checked out on the train. We see the machine all around us. Evil is not a mist. It's roman numerals carved into granite brownstone. The Pentagon is evil. D.C. is evil. Hollywood is evil. Courts and corporations are walking corpses ruling our world. Evil is their sanctioned choreography. There's nothing conscious about it. Evil is the antithesis of consciousness.

Joe Rogan's show *Fear Factor* was a televised welcome to the machine. It traded contestant's dignity for financial reward. Joe showcased evil as a tool for liberation. He framed the body's repulsions as a handicap. Joe won his success through the secretion of his dignity. The more recruits that join him, the less shuddering he feels in his bones. Evil is a race to a lack of regret. It's a pyramid scheme of self-effacement. It tells everyone, "just do it." It evacuates the spirit's desire to live inside a body. Evil pries the self out of its can with the energy of money, the power of influence, and the illusion of victory. The fallen corpses watching at home laugh with Joe because they feel less alone. Evil is the stuffed hyena gorging a child's throat. Evil is the evacuation of Holy. It's thoughtless piss from a drunkard watering an altar. It's a white virgin lace pressed into oily mud by a shitty boot. Evil is not the devil. It's human apathy amok with diplomatic immunity. Evil is the ease with which we turn our backs on God.

The moment we dehumanize evil we overcome. We become larger than it ever could be. Hollywood needs you thinking evil is alive and breathing. It makes you think it's anywhere but here while it spoons you on the couch. It intimidates your life-force and puts you on your knees. Rise up. God has knighted you a sovereign man. You are his burning sword forged in red plasma.

CHAPTER TWENTY-SIX
CNN is the Government

CNN's Veronica Stracqualursi is proud of herself. She gained an estimated $640 for writing an article claiming the state of Tennessee is requiring a law to honor the KKK. The story she wrote was about an American war veteran. She doesn't call him a veteran. She doesn't even show his picture. Instead, her story leads with, "KKK" and a full-color video of a burning cross. CNN pays Veronica to inject racial content to disparage the country by region. Veronica probably doesn't like disparaging people for a living. Her and CNN have a symbiotic relationship. She doles out racial hate from a cubicle in exchange for a mortgage. Airports are required by contract to play Veronica's work. She is in the unique position of owning a captive audience. Her demographic of travelers stretches across the country. Why? Why does Veronica get a captive audience? The answer is the government.

The media is the government. We the people don't know it yet. The media controls more than just our thoughts. The media steers our tongues down its river. As of today, the "R" word, the "N" word, and the "J" word are unspeakable. In the state of Florida, you can go to jail for questioning Israel on campus. These rapids are the workings of Veronica and CNN. What really happened in Tennessee was the state decided to honor a hero. A local America Veteran of the Civil War, and one of the first ambassadors of race relations.

Union General William Sherman called Nathan Bedford Forrest "the most remarkable man our Civil War produced." On July 5,

The Technology of Belief

1875, Nathan Bedford Forrest became the first white man honored to speak before what we now call the NAACP. He delivered this speech on the fairground barbecue outside of Memphis, Tennessee:

"Ladies and Gentlemen, I accept the flowers as a memento of reconciliation between the white and colored races of the Southern states. I accept it more particularly as it comes from a colored lady, for if there is anyone on God's earth who loves the ladies I believe it is myself. (Immense applause and laughter.) I came here with the jeers of some white people, who think that I am doing wrong. I believe I can exert some influence, and do much to assist the people in strengthening fraternal relations, and shall do all in my power to elevate every man, to depress none.
(Applause.)
I want to elevate you to take positions in law offices, in stores, on farms, and wherever you are capable of going. I have not said anything about politics today. I don't propose to say anything about politics. You have a right to elect whom you please; vote for the man you think best, and I think, when that is done, you and I are freemen. Do as you consider right and honest in electing men for office. I did not come here to make you a long speech, although invited to do so by you. I am not much of a speaker, and my business prevented me from preparing myself. I came to meet you as friends, and welcome you to the white people. I want you to come nearer to us. When I can serve you I will do so. We have but one flag, one country; let us stand together. We may differ in color, but not in sentiment. Many things have been said about me which are wrong, and which white and black persons here, who stood by me through the war, can contradict. Go to work, be industrious, live honestly and act truly, and when you are oppressed I'll come to your relief. I thank you, ladies and gentlemen, for this opportunity you have afforded me to be with you, and to assure you that I am with you in heart and in hand." (Prolonged applause.)

Julius Caesar supported slavery. So, too, did Cleopatra and Rome. Jesus, and his disciples preached support for a government run on slavery. Slavery will be around as long as men struggle with sovereignty. Slavery is training wheels. Slavery has always been symbiotic and codependent. It has nothing to do with race or culture. It's how a crawling baby learns to walk. Half the babies want to condemn everyone for trying. The other half want to help us stand. All of us are slaves to media. Even the ones who think they're not. Media is more powerful than friendship, self-love, written history, or logic. That's why media is government.

People pretend a man is forced into slavery when really he is coaxed. Slaves walk where they're told. Slavery is more voluntary than we are comfortable admitting. You know what they called an early American slave who said no? He was a Seminole. He was a Cherokee.

CNN is food court propaganda in the mall. For a network constantly concerned about racism they missed an opportunity to report on what has to happen for it to improve. Instead, they push Veronica out into the crowd holding a tray of fried dead chicken skewered on toothpicks for anyone to swallow. "It's free," she says. We're all stuck in the airport forced to watch her push samples. The smell of it gets deep in our clothes.

In the month of July, Veronica tweeted nine times for thirty-one likes. Five of her last seven stories focus on racial tension and division. Does this sound like a competitive news strategy for CNN? Or does it reveal a corporation managing the American psyche? The business of news is the collecting of eyeballs in jars. True competition would've sent Veronica back to the kitchen to fry something better. They don't. This is not about money for them. This is not about truth. This is not about ratings. What other motivation is there but dissonance?

Why would CNN want us divided? The first step to mind control is breaking the target. This is what a rancher does to cattle. It explains the motives of keeping Veronica despite her performance. No one but the media holds more control over our thoughts and our tongues. Media is the only source of food inside the terminal gates. CNN injects shame like ink into a tank so we start competing for who sees better. CNN will call the NAACP racist if it has to. These people want you on your knees begging for virtue. This is how you manage a human farm. You have to geld the spirit of everyone.

Veronica is here to reward you for flogging your ego. She creates a taste for emotional masochism. Terms like "white privilege," "toxic masculinity," and "white nationalist" are tools of her trade. If you think your genitals, melanin, or political affiliation is a source of shame, you are living in a syringe of self-loathing. This narcissism will manifest with you seeing the outer world as racist, sexist, or xenophobic, too. This is the technology of belief.

The prana economy rules the world. Government needs us to believe in the strength of its fence. It needs tension, scandal, intrigue, and redemption to keep our attention. Our life-force pours into the government like a wet fuel. CNN constricts approval of Trump so it can let it leak when needed. Out of Veronica's last seven stories, one stands out as positive. It's the headline that paints Trump as powerful when he talks about the moon. Keeping the country together requires beliefs like NASA. It creates the scaffolding of trust and authority required to keep the illusion. We give this machine 40% of our life-force. They tell us they landed on the moon in return. This is how the prana economy works. It requires our passionate belief to keep us turning the big wheel. They need a weak ego to enforce their program.

NASA's been struggling to hold their illusion lately. Trump is trying to repair it with SpaceForce. The Smithsonian broadcasting a Saturn rocket on the Washington Monument is alchemy. Trump the magician fusing patriotism with the space program. He makes it as expensive as possible to doubt the moon landing. To doubt NASA is to doubt America. This makes the act of questioning unpatriotic. These are the eddies and boulders placed in the stream. This is how you keep 250 million voters glued to two parties. Unite them under something big and claim ownership. The moon is the biggest, most valuable prana real estate in the world. Everyone sees the moon. Everyone gives the moon life-force. NASA claims it by calling it a rock they stepped on a long time ago. Buzz Aldrin even pee'd on it like a dog. The moon is America now. Conquering the mind's eye is the purpose of government. There is nothing nefarious going on here. Even the pharaohs of Egypt were thought to raise the sun. This is how you rule a hive. You fill it with propaganda you control.

If you want to convince people you can walk on water you tell them you ran to Hawaii. Overstretching installs a belief through compromise. Knowing NASA is a fraud still implants the belief the moon is walkable. That still counts as a mission accomplished for the godless council ruling the country's mind. Love yourself completely. Children are watching. We don't need more of them growing up in a world of lies.

CHAPTER TWENTY-SEVEN
The Prana Economy

Compulsory education is regurgitation. On a Tuesday in September, a teacher in Sarasota leads children casting a spell for President Bush. Their prana pulls down buildings in New York City. The teacher opens the ceremony. "Get ready!" "Kite." "Yes, Kite. Get ready!" "Hit." "Yes, Hit. Get ready!" "Steel." "Yes, Steel. Get ready!" "Plane." "Yes, Plane. Get ready!" "Must." When researching Hillary, I found a voodoo ritual where a priestess regurgitates black gumbo for the congregation. The disciples eat her potion thinking it gains them magical powers. But the gumbo is a lifelong possession. Like the priestess, teachers regurgitate spells to children. Our children swallow them for the powers of reading and comprehension. The installation of language forges a permanent bridge for possession. For the rest of our lives, sound enters the mind imbued with meaning. We are hijacked by language as it steers the pictures that form in the eye. We manipulate each other with the spoken sigils of alphabetics. All of us are possessing one another with the spell of language.

None of us need words to communicate. I can reach you deeper with my vibration. We need no schooling to comprehend. All we need is the desire.

The lowest among us is the atheist. These human creatures are the spiritually amputated. A cell's motor protein walking down a chain is not the stuff of entropy. Three minutes after Dr. Scientism insists

The Technology of Belief

the universe came from nothing, he tells us matter can never be created or destroyed. We are surrounded by meat machines of stupidity. If you feel I am hard on atheists, you're paying attention. Atheists are pissing in the pool as they drink at the bar. They carry themselves like the octopus dumping warm ink under a smile. May our cruelty be therapeutic. They've been here long enough, and clean water is better for everyone. I know we can turn zombies back into people. It requires a sharp stick and the willingness to make them uncomfortable. This is the energy work of prana. The spirit can grow back like a lizard's tail. In the words of Nathan Bedford Forrest, "Keep pushing boys! We got the skeer'd in 'em now."

When someone tells you the President is racist, ask them about Building Seven. The silence you hear is their endocrine system missing a gear. In one question, you show them how ridiculous it is to be moved by words from a man's tongue they already hate. Show them the true reality of government like a flasher opening his trench coat on the street. If you believe you can't change someone's mind or you can't make a horse drink, get out of the cavalry. We are Jedi alchemists on horseback. If the government can change a mind, so can we. This is the life work of prana. We transmute ourselves through the belief we can be better.

Words are the voice box enlisted. Our words move mountains and open gates. Don't cheapen your arrows by blunting the tip. Keep your tongue a taut bow as the mighty sky hunter. Sarcasm is a lie with no backbone. Disengage from the language of the hyena. Hyenas are beasts in the fluorescent hallway of the machine. They've followed so many orders they cackle at the lifeforms stuck in the system. They spit through holes in plexiglass, "I just work here." Your satisfaction is none of their business. Their eyes call you incredulous for expecting better. Don't join this pack. Honor the true captain of your will.

Parthenia is the energy of virginity built up over time. The altar on a mantle is charged by this kind of prana. When it's dusty, the energy is low, so you give it attention. You pour into it your awareness as you pinch away the cobwebs. The prana economy repays you the following afternoon. You left a gift for your self in the future. This is an act of self-love. Every action is an economic exchange of prana. Every moment is an investment or a withdrawal. Learn to see the receipt of every transaction you make.

Washington, D.C. is the psychopathic capital of America. We have to decentralize our consent to cut its supply. We can do this by changing the subject of government. Right now, as you read, someone with no birth certificate is roaming the clouds past the ice wall in an airship. He is a freeman. He has a friend in a submarine playing trombone for whales. These men don't play in our games of hunger. Society calls them pirates for not participating.

They are waiting for us to join them. Every day we feed the lie, we miss the golden dawn. Right now, our kids buy shrink-wrap firewood at the gas station. In every breath, you are reborn. Make eye contact with your mantle. Resurrect the world right now with the deep breath of intention.

The ego is your life-force. They shame the ego so you'll stop protecting it. They call prana Hades, so you turn your head and run. The prana economy is happening beneath our feet regardless of denial. We are urge equations firing a limbic solution aimed by the neocortex. We are energy, texting the universe through our skin. We are the synaptic passion of electric bubbles in the belly of a giant turtle. This whole place is plasma and belief from the rocks and the clouds to the thoughts and their passion. The prana economy does not distinguish good from bad. If you hate Trump or if you love Trump, you donate the same energy. We have to change our belief in government.

The definition of truth is engagement. What you engage is what you believe and what you believe is what you will call truth. This is why there's no difference between hating or loving a President. Engagement is a prana contract and a life-force subscription.

What is the true meaning of Hades? Is not the cleanest water drawn from the deepest well? Our skin, like grass, is a facade. We barter with coins in public to buy what we need in private. The prana economy is the underbelly of conversation. Beneath the rational, there is a black market where we buy and sell belief. We sell ourselves to buy ourselves back later. We become desperate to make a deal and ignore the mark of the beast.

Prana is the currency of life-force. Life-force is the destiny of utterance. The utterance is the spell of I am. Prana is its tendency to exist in the underworld. The world was created in ten utterances. Your destiny is a ticker tape dispersed by the vocal cords. We were born to make noise. Some rattle more than hum.

The Technology of Belief

To sense things is to unwrap their prana. Perception is the acceptance of photons. To understand each other is to validate and energize the heart. We behold each other when we talk. We exchange prana in the communion. We welcome someone into our senses and form an asynchronous agreement for life-force. What is said, and what is exchanged are rarely the same. Hades is the underworld we don't acknowledge.

Belief is a finite resource. Like time, you are limited to a daily allowance. When you believe in government, 40% of the belief in yourself is exported. We subscribe to beliefs with our attention. Attention is obedience. What you attend is what you believe.

The software of existence can be coded in two lines. Line 10, I Am. Line 20, Go to Line 10. There is power in a looping belief. Look at how they lie to catch it. They tell you it's all a vacuum. They materialize everything alive.

Photons carry agenda. Light is an idea turned to song. Magic always builds a fire when destiny agrees. Prana is dispersed through the world like berries on branches. All of our world is sprouting with prana. All of it charged with intent. The trees intend to breathe. The rocks intend to rest. The fruit intends to ripe and then fall. Some spells last longer than others. Granite, for example, is cast by the sturdiest magician.

Legend was a young and tall Creek magician. He gathered reeds between his fingers as he carved his way through the thick furry grass of what we now call Georgia. He formed a bundle of reeds into the shape of lips as he walked the forest perimeter.

At dusk, he split a cattail vertically, revealing it's puffy white center. His thumb smeared amber sap from a Cedar into the fluffy lips of the nest. He placed his sculpture into a ring of stone to perform his alchemy. He pushed a black pebble of char into the chamber of his rifle and cocked the hammer. Legend unleashed the trigger as flint cast its spell. A passionate orange ember the size of a pea ignites the barrel's crucible. Legend pours his fresh lava into the nest of reeds. The ritual of fire is born. The sap melts into the pristine white cattail and erupts in sparks. Like a dandelion exploding in the wind, the nest of reeds is forged in sparkling napalm. The spell unfolds naturally as the magician watches his incantation. Prana is the tinder of Prometheus. We plant our will in a circle and watch it unfold. Forget everything you've been told.

Remember everything you experience. You are a sacred witness embedded in life's tapestry.

Consciousness is a human ghost dance. Stories are the possession of memory. Every sentence you read echoes inside you the rest of your life. Thoughts are an invitation to trance. We are haunted by the spores of etheric genies disguised as thoughts. Bad or good, we banish demons from our circle like dust with a broom. We are the masters of the carpet.

A powerful magic is a falling orchid caught in a patient palm. It's charged with the prana of rescue from the fall. The flower is imbued with the energy of salvation. The orchid is made more precious by the alchemist's tender wait. The prana is co-created between them in a single timeline.

CHAPTER TWENTY-EIGHT
Government is Mafia

Under mind control, people think the mafia is rare. They think it runs underground or operates illegally. But a mafia is just a loose collection of conspirators with a theory. And a conspiracy theory is when two or more criminals in the government have a conversation about breaking a law. Society insists conspirators would never infiltrate the government. The term "conspiracy theory" is a mind control spell. It's one of the blackest we've seen inserted into the collective. Conspiracy theory is a drenched wet blanket over a kindled fire.

Mafia is not an anomaly. The government holds a monopoly on violence. The mafia holds a monopoly on crime. Both organizations are symbiotic. Both have structures managed and built from the same skeleton. Both grew exponentially under Prohibition. Both shrunk in hunger when it was lifted. If you constrict someone's breathing you can guarantee the first thing they'll want to do is breath. This makes us all predictable. People will never understand how easy it is to program a population. The difference between the mafia and the government is our belief in the sanctity of badges.

Government and mafia are corporations functioning as open secret societies. Under a chain-of-command, these systems insulate the organization from the outside. Omertà is the mafia oath of silence while under interrogation. Classified information is the government's oath under scrutiny. When a crew of boys is arrested by the policeman, their loyalty is forged in the back of the squad car. The government and the mafia are machines that feed each other. They place the organizations survival higher than their own. We have been dethroned by a machine made of humans that trades violence for gold and brass. The mafia does it for money. The military does it for valor.

The law, military, and mafia recruit from broken homes. The recruits find comfort inside the machine. They are baptized in death rituals and raised above the law. They operate inside a slippery bubble of morality. These men kill as part of their job. The most satanic phrase ever spoken is, "It's nothing personal."

After Prohibition, the mafia and the government merged. They consolidated resources and took over the human farm of America. Army and Navy units began grooming private squadrons of international gangsters. The Central Intelligence Agency began as a secret society inside the Army. In Kay Griggs 1998 interview on Satanism in the military, she outlined the initiation into the Brotherhood required a ritual killing on domestic soil. Just like the mafia, the machine of military and government feeds on human sacrifice. The recruit takes the life of his target. The machine takes the life of the shooter. Being decorated as a hero or a wise-guy is a death ritual. We treat the military as the golden child and the mafia as the scapegoat. Our lack of sovereignty is the only true cause to blame.

There is nothing rare about the mafia. The Boston mob has shipped weapons to Ireland under the cloak of the American military. Zionists from Germany used the military to launder money. The Brooklyn mob, the New Jersey mob, and the Dixie mob in Louisiana all have had their tentacles in the armed forces. Former Los Angeles FBI director Ted Gunderson confirms these reports and goes on to outline the mafia's role in law enforcement. "Who you gunna tell?" is a mind control campaign of gang intimidation. The symbology on shields and uniforms reveals the coverage of these campaigns. The mafia hunts in broad daylight on our roads and in our squad cars. We believe their badges and medals are incorruptible. We do this because we want to believe. Our scrutiny's calories are better spent on the mortgage.

The Great Seal of the United States officially has two sides. The obverse, or front, is the eagle in the coat of arms we see with the President. On the Great Seal's reverse is a pyramid with the Eye of Providence. We see the eye on the back of what we call money.

The Technology of Belief

When the President speaks in front of the Seal, he is speaking directly to the pyramid on the back. He makes promises to the men behind the curtain. There are three mottos on the Great Seal of the United States. Our belief in Washington supports every one of them. How is this not a New World Order color by numbers?

1.) E Pluribus Unum. Out of many, one.
   2.) Annuit cœptis. Things undertaken.
   3.) Novus ordo seclorum. New order of the ages.

When I joined the Navy, my recruiter told us we should bring all the cigarettes, candy, and magazines we could pack. I arrived in San Diego for basic training early in the morning. When our group got off the bus we were told to separate our donated contraband into three piles. This was the first lie. The upper ranks chuckled as they regained their lost dignity when they wore our shoes. Lies are the glue for camaraderie. The individual is powerless under the golem of seniority.

Studies show 100% of people thrown into a pit will immediately desire salvation. Ascension programs always start from a fall. From the moment I got off the bus, I was thirsting for levity. They shaved my head. They changed my clothes. They took my name. I would do most anything to stay above the dark water. Pits are gatekeepers for ascension programs and victim predictability. You fall in their enlistment. You rise when you obey. The experience trauma bonds you to the corners of the machine. You are promoted in the military. You are made in the mafia. You are elected to Congress.

Gunderson claimed the first World Trade Center bombings in 1993 were FBI mafia operations. They link Bush to the Clintons via the Arkansas drug trafficking business hidden by the Rose Law Firm. Again, elite military working with the regional mafia. Mobs aren't anomalies, they are scaffolding. Through the mafia, the cabal feeds the spider. Brotherhood JAG officers infiltrate local courts as judges and political appointees. Michael Aquino is not an anomaly. Griggs says most full bird Colonel's are promoted through blackmail and depravity rituals. A recruited, poor, fatherless young man is evil's wet dream. The cabal doesn't need flags, religion, patriotism, or even a common language to thrive. It only needs vulnerable broken young men to do something they are ashamed of for the promise of a pedestal. This is the recipe for vampires.

The Brotherhood is everywhere. Their symbols do more than intimidate kids. They reinforce obedience. It has to be out in the open for it to work. Missing children is the transmission of the cabal. Without human sex trafficking, the mafia goes nowhere. A concerned witness in Missouri isn't allowed to track a missing child by photo for 30 days. In 30 days the child is gone. This is not an anomaly, it's a nationwide feature. Do you see the world wide web? They dangle it in our face and in the media. The spider is so huge the fear makes us blind and oblivious. We think it's all random anomalies but the machine is perfect. This is the anatomy of a cabal. It's time to make sleepy neighbors uncomfortable. It's time to put a tax on apathy and denial. Epstein is a door. The zeitgeist will only be open for so long. Be smart with your words. You don't need to convince anyone. You only need to make it uncomfortable to ignore.

The Anatomy of Satanism in the military can be summed up in one quote from the movie 'A Few Good Men.' Lt. Kendrick (Kiefer Sutherland) explains his views on sovereignty to a court, "The only proper authorities I'm aware of are my commanding officer, Colonel Nathan R. Jessup, and the Lord our God." The Lieutenant does not grant himself a position in the chain of command. The military industrial complex's chain of command usurps autonomy. Broken families makes the harvest possible.

The intercontinental mafia is an army of existentialists. They read Clausewitz, Neitzche, Sartre, and Montesquieu. They are moral relativists. According to Griggs, Brotherhood members would write about these men as a senior thesis to pad their resumes. Moral relativism in the military is summed up Clausewitz, "War is the continuation of politics by other means."

"In all my career at ATF, the people I put in jail have more honor than the top administration in this organization." – Agent Bob Hoffman

The Waco, Texas Massacre was a pillage ritual used to cover the murders of four Clinton witnesses. The evidence for the raid at Waco was an unsubstantiated rumor of a $200 unpaid tax on a single automatic rifle. The ATF came into Waco broadcasting the sounds of rabbits being tortured over loudspeakers during their siege. This was our government and they planned this ritual for months. In the Linda Thompson documentary, *Waco: The Big Lie*, at the 12:46 mark, you see one ATF agent kill three of Bill Clinton's former bodyguards after they enter a window from the roof of Mount Caramel. A fourth Clinton bodyguard was killed in the parking lot before the siege began.

Timothy McVeigh listed Waco as a contributing factor for his attack in Oklahoma City. His execution was carried out four times faster than any other prisoner in federal history. In 2002, Gore Vidal wrote McVeigh "told us why he did it at eloquent length. But our rulers and their media preferred to depict him as a sadistic, crazed monster ... who had done it for the kicks:" The Clintons are only one of the many mobs running America. Arkansas crime was not their creation. Infrastructure is required for the giant spider to feed. Their hit list is well over 200 domestic murders. Sixteen deaths are from plane crashes alone. The Waco massacre was trivial.

I've only told you about four American mobs and we haven't even mentioned Chicago. Obama's first two elections were won on technicalities and disqualifications from inside the party. In Chicago, the mob united into a monolithic force called the Outfit. Mob brotherhood permeates America deep in its shorts. Shitbags like Sheriff Israel are a standard feature in cities across America. All of them thrive under the singular watchful eye of the FBI. How's that working out for you? How's that working out for the spider?

From the drug cargo ships of J.P. Morgan to the cocaine submarines seized by the coast guard, the mafia is an independent state living inside our country. False flags are the satanic machine priming the system for another season. Terror and intimidation are the melancholy Santa ringing a bell outside the mall. These uneasy feelings fill up their coffers. We have no idea what happens to that money once it drops. Satanic evacuation requires a constant stream of reliable injustice. America has been drowning in its spinal fluid for centuries. The Brotherhood has permeated every nook and cranny of our military and our justice department. They can turn any of us against any of us with one call.

We shed ourselves to join Hollywood and the military. We sign up to be raped and rebuilt with a new appetite and compass. We mark success with silver medals and golden idols. We are cheered for our acts of self-destruction. The Military Industrial Complex and Hollywood eat our flesh to stay alive. Like Hollywood, the tree of liberty is covered by chameleons of red, white, and blue. Satanism promotes a loyalty to colors. It worships our ability to wear a uniform and focus on the job.

The Brotherhood mob is decentralized and distributed. This makes the spider web invisible to the eye. We only see a creature here or an insect there. We are in complete denial of the mother living underground. She lays new eggs each day inside our trust. This is the anatomy of Satanism in government. It has henchmen, agents, directors, and tycoons. It goes higher than the president of a country. It goes higher than Kissinger but we'll never hear those names. Satanism outranks us all. As long as we fear ourselves, we will continue to work for an international human cartel of evil we call the government. We will protect their assets in broad daylight with naive ignorance at the hypnotic sight of badges and medals.

CHAPTER TWENTY-NINE
The Wasp and the Caterpillar

For fifty years, every President has promised America we would return to the moon. For fifty years, every President has given NASA a mandate that's bold and grand. On July 20th, 2019, Vice President Mike Pence gave a speech to the Kennedy Space Center. He said what's been said before and raised it a gender. Pence said a woman will walk on the moon in five years. For fifty years, every President has promised something amazing to be delivered shortly after they leave office. Pence nailed the timing perfectly. How quickly each election makes us forget.

Most of us accept the Apollo Program was a fraud. Most of us know the government is in the business of lies and persuasion. But people still follow the loyalty of their chemicals. Those chemicals are the hook. International pedophilia is the latest moon landing and QAnon is the new Captain America. Admission to the narrative requires one small agreement, "we're saving Israel for last."

While Syrians, Iranians, and Palestinians are fighting for their sovereignty and lives, American politicians unanimously agree on Israel. On June 28th, a dozen American F-22 Raptors return to the Persian Gulf. America's war powers remain subdued by dual-citizens. On 9/11, an Air Force JASSM plowed into the walls of the Pentagon as we pretended it was a plane. A Saudi prince loses power as gunfire spills into a Vegas crowd under the eye of a black pyramid. The 9/11 venom attacked the country's immune system from within. We are a helpless caterpillar as a new world order gets closer. We are comfortably numb from the venom burning in our backs.

If QAnon was headquartered on a Florida campus they'd have an excuse for not talking about Mossad. It's illegal there to mention a Jewish world conspiracy. The figure called Q doesn't talk about the first amendment. The first two words of "Saving Israel for last" are "Saving Israel." Every administration only has so much time. If another Mueller feather was placed on the table the whole game would stall. Articles of impeachment are the sideshow excusing the lack of movement. If QAnon was legitimately fighting for America it would be pointing to Mossad. They would leverage Epstein with everything they've got. They don't. They won't. They can't. Elite professionals don't miss the perfect shot. Nor would they shoot their own foot. Epstein doesn't hurt Israel. It hurts the Clinton Foundation. She's the witch QAnon will give us to burn.

Mossad is the CIA of Israel. It's an international belief cartel living in our government. It's a tapeworm with diplomatic immunity that we pay to chew our legs off. Its logo is the tabernacle for the Holy of Holies. Its motto is "by way of deception." Mossad is a weaponized golden child on a belief pedestal. Its power comes from the worship of a lie. That lie is the belief God has a chosen people. As the truth comes out, the corruption of Abraham would finally die. It would pull back the curtain of the Tabernacle and reveal the temple of the golden narcissist. Chosen people worship themselves not God.

Mossad steers media and party. It controls Hollywood and Disney. It controls Congress. It touches billionaires in Ohio. It touches a child in a dentist chair in the Virgin Islands. If we hold our tongues against it we lose the chance to speak again. There are brutal blood lords squeezed into human bodies running the kitchens. The genes of Mossad have been tailored with ancestral brutality. Mossad has forged Saturnian inversion into a global domination policy. They inject human sacrifice, burnt human offerings, genital mutilation, genocide, and blackmail into the veins of the world while they thrive. Globalism is the virus of cannibalism unleashed on the world. We are building a pyramid for a New World Order. We are preparing ourselves for the volcano.

It's hard to convey just how deep we are in the belly of this spider. It's impossible to point at your world while you're living inside. If I point to Trump I am unpatriotic. If I point to Israel, I am antisemitic. If I point to QAnon, I am disinformation. If I point to Mossad, I am excusing the Vatican. The wasp plants its eggs in the back of a caterpillar and befriends it long enough to hatch. It tells the caterpillar soon it will be able to fly. We are ruled by the venom of salvation. The rapture from our backs won't be from sprouting wings. Soon the eggs will hatch and we'll feel the truth crawl. We are still in our dreamy haze of hypnosis. For now, we are the happy worm on a mushroom smoking their lies.

Andrew Breitbart is famous for "showing how the Internet could be used to route around information bottlenecks imposed by official spokesmen and legacy news outlets." Andrew Breitbart was a brilliant caterpillar. His vessel was seized with the venom of secret information. In June 2011, Breitbart broke the story that congressman Anthony Weiner was sending young women revealing photographs of himself. Those photos weren't stolen. They were leaked like venom. They were held in a safe and traded on the open market. All of us are ruled by our chemicals. Andrew Breitbart was a mover who wanted a canvas. Our desire makes us an asset if we feel empty without it.

"I'm glad I've become a journalist because I'd like to fight on behalf of the Israeli people." – Andrew Breitbart.

Mossad hatched their eggs inside Breitbart. Andrew never had a clue. On the night of March 1, 2012, Breitbart collapsed suddenly while walking in Brentwood. He was rushed to Ronald Reagan UCLA Medical Center and pronounced dead just after midnight. An autopsy showed he died of a heart attack. He was 43 years old. Larry Solov, Breitbart's childhood friend, served as general counsel for the company since 2007. Solov boasted the Breitbart organization was conceived in Israel. In 2007, he published a photo of him, Netanyahu, and Andrew as evidence Israel was behind them from the beginning. After Andrew's death, Solov became president of Breitbart News the same day Steve Bannon was named Executive Chairman. In a December 11, 2016 interview with 60 Minutes, Benjamin Netanyahu said that he didn't know Steve Bannon. The Israeli PM told 60 Minutes he's not concerned about charges of anti-Semitism among Trump's supporters and trusts Trump will set the tone. Eleven days after Andrew's death, the

QAnon team was ready to strike. Trump tweeted:

"When I was 18, people called me Donald Trump. When he was 18, BarackObama was Barry Soweto. Weird." - Donald Trump

On July 19th, 2016, Trump became the official party nominee. One month later, Trump hired Steve Bannon. Three months later, Bannon and Trump won the election. Bannon worked as Chief Strategist in the Trump administration for one year and one day. His exit was flawlessly timed as Trump tossed him like a professional wrestler out of the ring. Bannon and Trump manufactured distance for the public to eat. Trump said "Sloppy Steve" had "lost his mind." Bannon left the White House on August 18, 2017. QAnon's first post came ten weeks later. Bannon has always been a proud Christian Zionist. The pack moves quickly under the stealth of night. When Bannon went dark, Q's sun rose in the light. Bannon stayed invisible until 2018. "Bannon expressed regret over his delayed response, declared his 'unwavering' support for Trump and his agenda." If the people knew Q was Bannon, Captain America would fall off his motorcycle.

Before 7 AM, Aug 8, 2018, a reporter caught Steve Bannon entering Epstein's Upper East Side mansion. The following day, QAnon posts "Showtime!" and named the Clinton Foundation linked in human trafficking. Why would Bannon pay Epstein a personal visit the day before this announcement? It reminds me of wolves warning wolves.

There were two wolves in the pen when I got there. The white one came up to my waist. The grey one was taller with sharp rugged shoulders. They carried themselves in a nimble trot and left puffy clouds of breath for the moon to eat. You could hear them howl for miles. The pose in their eyes was silent and vorpal. You tame a wolf the same way you tame a sheep. You mirror. You demonstrate alliance. You alter the vector. In hypnotic programming, this is the art of pacing. People love the taste of finishing each other's sentences. People feel safety in the camaraderie of familiar.

Democrats and Republicans are in complete agreement on Israeli aid, Palestinian genocide, antisemitism laws, and dual-partnership with Israel. If you want to see our ruler's fangs, look at what both parties have in common. America First is impossible with Israel.

The Technology of Belief

Our President is entrenched on Israel's lap and suckling at her teet. Every day it becomes harder to say it isn't true. Zionism has killed our sovereignty. Donald Trump fails America by fighting for it. Donald Trump can't believe in America First with his loyalty abroad. Polygamy is just as destructive when you apply it to nations. We can not be Israel's keeper or its accomplice.

One day the Christians created a code of behavior for non-Christians. They built non-Christians an outer church under a central religious authority and outlined when non-Christians could be put to death for infractions. The Christians accused the non-Christians of persecution. They demanded reparations through special hate laws and billions in emotional blackmail.

If the above passage sounds impossible, change the word "Christians" to "Jews" and read it again. The Talmud is an oral tradition of mind control under a centralized Abrahamic Lord. See the Old Testament for acts of sacrifice and devotion used as gauges for loyalty.

The caterpillar is paralyzed by the wasp. The caterpillar is not in pain. The caterpillar is under a spell. The Florida antisemitism laws against free speech are the venom's paralysis over the tongue. The Star of Remphan on every law enforcement badge and squad car in Hollywood is a form of priming. We ignore how dangerous religious narcissism can be to our environment. Evil comes with chemtrails of apathy and silence. Our grandchildren peek back from the future as we feed ourselves to the machine. They see us cheer for a modern King Cyrus. We make excuses for propaganda that tells us things are getting better. Even if they are, the venom of anticipation has been winning for a century.

We are subdued by a wasp who waits for the eggs to hatch. If "Israel is last" why are they hailing Trump on coins as their champion? Does it sound plausible the 102-year-old Balfour Plan was defeated in a general election. With the help of Zionism? Or is Mossad being deceptive again? The tools of Mossad are blackmail, pedophilia, and Hollywood. Epstein was versed in all three.

It's hard to look through the smoke of a holocaust. That's why it's called a burnt offering. The story of Moses comes wrapped in an ancient blindfold. Our salvation is swaddled in spiritual codependency. We'd rather be rescued than get our feet wet.

The One enthroned in heaven laughs; the Lord scoffs at them. He rebukes them in his anger and terrifies them in his wrath, saying, "I have installed my king on Zion, my holy mountain." - Psalm 2:4-6

Friends, I haven't mentioned Jared Kushner. There's too much soup to sift in one sitting. We fail when we get too busy watching it all happen. It's too unbelievable to move from our present position. We are the baby under the finger of jingling keys. We squander our energy on anonymous theorists with nothing to lose. We get drunk on news that claims to be hiding around the corner. We are mesmerized by the sale of fermented wine from the prison toilet. My views on QAnon won't keep anyone high. Nor will they get you drunk. All I have is a sharp spoon I've been using to carve a hole through the wall. I can smell fresh moonlight. As the slaves of hope sleep the truth keeps us digging. America is outside these walls. We claw our way out and reach for her skin under the concrete.

As of this writing, none of the representatives in Washington investigating the Epstein case have heard or are aware of any Israeli intelligence connections between Epstein, our government, or Israel. The venom is working perfectly.

CHAPTER THIRTY
The Second Coming

From the time he was born they told him he was perfect. And so he was. No one's spine towered taller than Jared Kushner. He had been trained for the best by the worst. Charles Kushner, the father of Jared, is a lifelong friend of Benjamin Netanyahu and a convicted felon. The two of them would turn Jared the child into a messiah for globalism. Such is the life of a thoroughbred human. They're carved from marble for a purpose. His chemistry was tuned as early as age four. Netanyahu tells the press about the night he slept in Jared's bedroom. His tender story is a ritual anointment before the world. Jared will win every room with his essence. Everyone agrees he's the chosen of the chosen. Behind a slender skull, Jared discovers his aura is more valuable than he is. He's a raccoon curled up in the cold fireplace of an abandoned mansion.

A treasure chest is a mystery before it's opened. When eyes peak, the probability wave collapses into particles. An X on the map marks the spot where we bury great expectations. Before the death of wonder, anticipation was the knife pealing its trough through the water. Jared Kushner was the masthead on the ship of Zionism. He will lead his ancient crew wherever they turn the wheel. They polish his wooden boobies and paint his hair with pride. He's a figurehead crucified in the wind on the beak of the "Second Coming."

On a Captain's orders, the media raised Kushner up from the forecastle. They were generous with their trappings. Fox heralded him a genius for selling red hats on Facebook. Hillary supporter Eric Schmidt said, "Jared Kushner is the biggest surprise of the

2016 election." Peter Thiel said, "Jared was effectively the chief operating officer." A resigning Nicki Haley called him, "a hidden genius no one understands." The golden child is raised up from both sides of the aisle and placed on the golden pedestal. He is the media's Ark of the Covenant. He rides the glory seat as a living sacrifice. Jared's purpose is to be the center of the world.

In Hebrew, the word Messiah means "Anointed One." Admiration is the modern version of Holy Oil. Kushner was anointed from all sides by a kingdom made from money and cameras. He has a part to play to repay his debt like Perseus. Freewill was removed from his frame a long time ago. Trumps says Kushner is the only one who could broker peace in the Middle East. Israel's defense minister, Avigdor Lieberman, said, "What we know, he's a really tough, smart guy, and we hope he will bring new energy to our region."

On January 12th, 2007, the most expensive building ever sold was bought by a 25-year-old Jared with a downpayment of $50 million. The 666 Fifth Avenue keys fell in his hands for $1.8 billion. Number one and eight are the sum of three sixes. The address and the price are harmonic to the magician. Jared was being anointed in numerology's oil. 666 is the kabbalistic number of the sun. It's also the number of man. Carbon is an element with six electrons, protons, and neutrons. Kabbalistic magicians keep the outer church from tapping into their magic. 666 is shrouded as the number of the devil and the mark of the beast. Fear is psychic fencing around the symbology. When Q says symbology will be their downfall. He doesn't say symbology is how they rose. 666 is the tuning fork of the Kushner messiah. His anointment gives everyone a vessel to hold their belief.

Jared lost $800 million on 666. The media doesn't mention that in their reporting. They protect the golden child from tarnish. CNN cleared Jared by saying his father, "pushed Jared to do the failed deal." Charles Kushner said, "We shy away from anything that could have a negative impact on Jared." The network anoints him with more generosity as they expose a plot against him with hints of antisemitism. CNN hates Trump but loves Kushner.

It's weird how Jared Kushner's dad paid two prostitutes $25k to seduce, record, and blackmail two witnesses. The US Attorney who nailed him was Chris Christie. Kushner is well-versed in the art of blackmail, extortion, and sex tapes. Anthony Weiner knows this kind of leverage all too well. So too, does Epstein and Clinton. Why did CNN miss the chance to print the headline, "Trump's in-law convicted of blackmail." That's a lot of jet fuel left in the sun to evaporate for no reason.

Charles Kushner got two years in prison for tax evasion, illegal campaign contributions, and witness tampering. The anti-Trump media could've made a Lifetime movie series out of it but they didn't. The Kushners have been anointed by the media. Nothing would stop the second coming.

"[The Kushner case was] one of the most loathsome, disgusting crimes that I prosecuted...and I was the U.S. attorney in New Jersey." - Chris Christie

When Israel and their media anoints a golden child something special happens. A hive of billionaires turns their needles in the same direction. Kushner becomes Polaris; a human Mecca for Zionism. And when billionaires turn, so too must the people's allegiance. Israeli American billionaire, Haim Saban ($3.4 Billion) was a heavy donor to Obama and Clinton. Saban's keynote address at his annual event features a conversation with himself and Jared Kushner. On stage, Saban pours prana onto Kushner as the bringer of a new era. The power of the messiah is the life-force of mass agreement. Belief is a thick stream of plasma with a charge. When it's ordained or anointed the effects are amplified exponentially.

All of us are vessels of belief's holy oil. It pours out with our engagements and attention. Time rubs it deep into the altar. Thiel ($2.5 Billion), Schmidt ($11 Billion), and Saban ($3 Billion) are vessels of prana revered in their respective congregations. What the billionaire believes, the flock adopts. The entire budget of the Republican and Democratic committees is less than 2% of Les Wexner's fortune ($7 Billion). Wexner is an owner of human horses like Epstein. He's not the only pedophile someone keeps in their stables. Wexner is a billionaire shielded by the forcefield of influence. To end pedophilia, we must withdraw our consent from their puppets and money.

We underestimate the power and influence of billionaires. In many ways, they are above government. They are formidable opponents when you measure their influence and firepower. George Soros ($26 Billion) has been banned from operating his NGO in his own native country. Corruption will call a One World Court into existence. A guilty billionaire on trial would drag us all into the machine while we cheered "justice." From NGO to NWO, we are cornered by billions of cones directing traffic through the toll.

There's just no fooling some people. Or so they believe. Smart people are blinded constantly being told how smart they are. Their deficit exposed would feel like a violent attack. This is the essence of the golden child syndrome. It is so blinding doctors can't fathom vaccines as dangerous. Engineers can't picture a NASA fraud. Lawyers lack the ability to scrutinize Building Seven. Golden children are abused on a pedestal. Scapegoats are neglected in the basement. Both suffer from abandonment enforced through isolation. These are self-inflicted technologies of belief.

Trump's victory hinges on a wall. Nothing helped that more than a migrant horde rushing the border. Our controllers wanted an archetypal King Cyrus. He will carve the way for the rebuilding the Third Temple. Just like Breitbart, Trump was made in Israel a long time ago. He was intentionally painted as the underdog no one thought could win. His victory in the White House was secured with the loins of his daughter. Her marriage to the golden child was a bloodline contract. Trump said, "If Kushner can't bring peace to the Middle East no one can." And so he will. Jared Kushner has been named the Moshiach.

"The law will go out from Zion, the word of the Lord from Jerusalem. He will judge between the nations and will settle disputes for many peoples. They will beat their swords into plowshares and their spears into pruning hooks. Nation will not take up sword against nation, nor will they train for war anymore."
- Isaiah 2:2-5

Julius Caesar painted his face red to ascend the temple of Jupiter. Trump wears orange make-up to channel an aura of gold. Like Cyrus, Trump will bring in a messiah for the Middle East. At least that's what they expect of him. There are eleven Holy Spices in the Tabernacle incense. For the first time in 2000 years, ten of them are burning. A virgin lamb was sacrificed in the presence of the

The Technology of Belief

Temple. The carbon-black ash of hate in the Middle East has been stoked for a ceremony of cleansing. Peace will reign as a golden child sounds his horn. In Rome, narratives like these were hatched by the families of Triumvirates. These were the informal governments within a formal government. Secret societies still work in the open.

As rehearsed, Marc Antony stood up in the crowd and said, "I wonder? Could it be that Julius Caesar is a God who walks among us?" The puddles of people in the courtyard stopped to look up in unison. They turned to Antony like a compass clutching pearls. The insertion of human divinity spread through everyone's mind like a tone. Caesar waited a strategic two minutes for the idea to simmer, "Now. Now." The grumbles parted as Caesar addressed the matter like a matador. "Marc Antony lays laurel where no man's feet may go. I assure you, by my mother's complaint, my feet were made from clay, not heaven. As was my cock that crowed that very first morning." The crowd burst into laughter at the joke. Caesar had stuffed his fingers into his armpits like a chicken and pretended to peck Marc Antony for his suggestion. These men commanded the vanity of laurels. Two years later, Julius would be hailed a god in his newly consecrated temple. His rise to Mount Olympus came from a well-crafted narrative. Caesar and Anthony were technicians of belief. They employed it with great profit. The elite have had two-thousand years to perfect this technology. Our engagement is proof its working.

CHAPTER THIRTY-ONE
The Capital of Punishment

The capital of punishment is ownership. We pull weeds from the ground we own. We protect our children with a spell of banishment to their room. How we treat life is complicated. Let's take a moment to lift up its sanctity. It seems the zeitgeist has stopped celebrating abortion as an instrument to kill fascism. Good. Let's use this momentum for a purpose. We mocked life for so long we forgot it's priceless. Two things happen in a system where everything has a price. One, we lose the meaning of priceless. Two, we gain the meaning of worthless. Putting a price on things is a spell we imbue in the aether. It sticks like goo to what we assign for the rest of our life. These grapes are $3.99. That racehorse is $7.2 Million. We traverse the matrix with a price gun tagging everything with a mark of the beast. We place our children for sale on its shelf. When I was fifteen-years-old, I was worth $4.25. We sell our essence by the hour. We pretend its sane to chase zeros and decimals.

A poorly managed state secretes dissonance on purpose. Angst is wet silk the spider spins into cream. Anti-fascism is the other set of legs from fascism. Corruption wraps us in a shroud while we choke each other out. We make our own anesthesia as it drips from an IV. Hope is not our friend; it's our drug dealer. There's a fine difference between it and faith. It's as hard to see as the thickness of paper.

We've reduced ourselves to scrolls. Were using a spider's money in a spider's court with words from its hairy mandible. Fangs assign their price to all of us eventually. She is the blackest of widows, and we never see her face.

The Technology of Belief

This entire operation is an energy farm. It's agitated from a laser pointer media. It shines a spotlight on a field of sheep huddled under sky's blanket. The flock is zombified from marauding coyotes dressed in wool. A blue donkey stands guard in welding goggles and galoshes. His sheep are starving for justice. The lack of it sprouts an addiction as they hunger for blood. They believe evil is hatched from the ribcage. The whole system has been wired for response. We pretend this place is broken as we snort hope behind the stairs. The flavor of fear's adrenalin marinates our meat for the owner's plate.

We are the energy of thought. The range of our beliefs puts limits on our thinking. If we insist our voice won't carry outside the tent, we never see the Aurora Borealis. It heard us whisper a phoenix last fall. It heard the eagle rise last night. We are enclosed in a cave with the trappings of mind. Stalactites are the petrified tears of a hungry sky. Under the blinders and bridle of society's reigns, we are hypnotized by the clomping of feet. They tell us we must be free since they keep moving. Git along, little dogies. We justify every step we take to feed the machine.

If prisons weren't greenhouses, we'd see justice. A jail is as simple as a dog kennel. Do dogs have an underground drug trade run by guards? Do reformed mutts come out with chips on their shoulders? Are beagles radicalized by the FBI? We become addicted to justice when it's inconsistent. We develop an unhealthy relationship with something we need to survive. The dissonance keeps us feeling worthless. And so we are.

Can a mother kill her baby with a reason? Can a father kill his son with one? Can a state kill a citizen with the same? We live in an insane society yet we insist on giving the warden a key. The price tag for a human is a barometer for how much we reject sovereignty. The vector of slavery is inversely congruent with the sanctity of life.

To rage against the machine is to feed it. It only starves when we change the subject. We want capital punishment because we were chemically trained for the taste. It comes with sinister changes to the role of the state. I don't need you to agree. I know many want the gallows hung higher. But we are getting worse at freedom, not better. It should be something we sip on after dinner.

If the state can take a life, what else can it take? If the state held no sovereignty over our lives, they would hold no sovereignty over our labor. We would have a unanimous government again. Like the people of the longhouse did for 500 years.

It's hard to tell the left abortion is wrong. It's hard to tell the right capital punishment is the same. Convincing either side to stop murder is nearly impossible. If a mother says, "I tried to punish him with no luck. So, I killed him." Is that okay? If not, then why is it ethical for the state to say the same? This is an exercise for our thoughts on sovereignty. We know the state is insane yet grant them authority like they're our parent. We give the power of life to a random committee of jurors. A committee is a corporation. We would call it inhumane if this decree came from a single person. We prefer the machine. We hold it in a higher esteem. Our legal system is a modern witch trial where everyone dresses nice. It feels far too modern to call it evil. There are only two sides in the quest for victory. To agree or not to agree. Victory is the answer.

The prime argument for capital punishment is deterrent. So how can we believe in a wall but not prison? Both are built on the same foundation of thinking. If banishment deters crime let us banish our felons in prison properly. Or should we shoot people at the border to be sure? Some already agree. Their thoughts should accompany abortion. Both take life on the ethics of utility. Wishful thinking decides a state would never execute the innocent. Psychotic thinking believes a few wrongful deaths every now and again are acceptable. We are our own worst slave masters. We spit on our sanctity every day.

Lately, I notice people wanting to disagree before I speak. By habit, I try and stop them, which is wrong. I am blessed by the foolishness I see in society. It reminds me of what a fool I can be. There are cracks in everyone's character. They're glazed with magnesium and cobalt and cast in fire. Our splintered veins are beautiful when we bake them in the oven. None of us need to agree. Conforming is fuel for machines. It crams living people inside it and separates them into two piles - those who agree and those who don't. Dichotomy is the essence of slavery. It's wisdom, not unity, we should be chasing. Unity is the nectar of zeitgeist. Wisdom is the nectar of logos.

To think and to believe is the mind breathing. We pull thoughts into our mind's lung. We exhale the belief. We harvest pollen from every decision. Decisions are a trance we play on the train while we

The Technology of Belief

travel. Thinking is the opposite of flow state. It's a goose who's stopped for directions. The moment we think our world is obscured by a holodeck. Like an overhead screen in a video game, it's helpful but not immersive. Immerse yourself in the game of toes and fingers. Find power and health by squeezing its fruit on the tongue. Carl Jung called thought logos and belief eros. Our decisions are arrows shot from the tension between mind and spleen. Our will is the cord stretched from the teeth of reason and imagination. We launch ourselves through space and time. Faith is the release of feathered vanes spun through a dynamo string.

CHAPTER THIRTY-TWO
The Two Towers

It's not the Jews. At least not exclusively. The evil we face is inversion. It's moral cannibalism and self-mutilation. Inversion is a contagious pathogen of the mind. It's a possession of the empty-hearted. It's a walking sucking chest wound. It turns humans into drones for a hive. Each finger sprouts another hand, and each hand sprouts five more fingers with five more palms. The fractal curse has infected Judaism, Catholicism, Christianity, Islam, science, fashion, government, mafia, military, and entertainment. A suicidal tone has been struck in the aether. It's a calling for the tower of Babel. What the pyramids of Egypt show are the power of devotion and obedience.

America suffers the same infection as Israel, England, and Rome. Our enemy is human evacuation. We see this manifest as psychopathy. The most dangerous beast in the jungle is the man with a hollow chest.

I always knew we'd talk about the Bible. I always knew it would make us uncomfortable. King James outlawed the spoken word by calling the tongue an instrument of evil. He declared a prohibition on the power of vocal cords. He declared magic a sin punishable by death. He said this before he published his Bible. In his book, *Daemonologie*, he said it plainly for his day.

"I mean by such kind of charms as commonly daft wives use, for healing of forspoken [bewitched] goods, for preserving them from evil eyes, by knitting ... or doing of such like innumerable things by

words, without applying anything meet to the part offended, as mediciners do." - Daemonologie

God never told you to find him in the Bible. People told you that. God talks to us, but he doesn't speak English. That would require a man to be trained first by other men to understand his tongue. God needs no middle man. God moves in the eros of your spleen. We feel him as the breeze during our tragedies or inspiration. I see calligraphy drip from his pen when it rains. I lay in my bed, and the moon tracks across the curtains. I see God's light poke through his vaulted firmament. God speaks in everything that moves and doesn't move. God is the spoken word, and King James outlawed us from using it.

The Bible describes a technology called LORD. This technology requires a belief you are broken. Wild horses are broken. May we look to the stars instead of man to see the scripture around us. We are holy submarines submerged in an ocean of atmosphere.

God is not found in a Nicene Creed. God did not sign his name in the Bible. The belief in Saints is a chain of command. Yesterday, someone told me it was okay for a state to take a life because it was in the Bible. Last month, someone said the Bible condoned taxes, so it is our duty to render them without question. People insist Palestinian genocide is impossible since they are God's chosen people. LORD technology is a magnet that bends our compass. It was designed to work that way.

Why would God lock his words in ancient greek and hide them in one geography and time? Why would he leave his most vital notes filtered through man's fingers? Why would he establish a chain of command through the Father, the Son, the saints, and the church? Why not speak directly to us? The Bible's foundation came from the Book of Moses. The same men that lie today have lied for centuries. 'veThey've gotten quite good at it because we are so prone to believe them. The Holy Spirit never whispered to any of us, "I wrote the Bible." It was always a company of men who insisted it was divine.

Salvation is a powerful trance. If you doubt, you are told to add more Bible. The answer is always in the book. The flaw is always in the reader. The scripture is not to be questioned. To question is to doubt. If you don't see God's grace in the pages, your heart may not be worthy. Read some more. Pray for guidance. The book is flawless.

We worship the Bible as a relic despite it having a commandment telling us not to. This is an example of Christian inversion. People insist they don't worship the Bible. But calling something flawless is an act of worship. Our devotion to its text is deeply ingrained obedience. If you don't like the answer, you should read quietly until you do. In Christianity, like Islam and Judaism, to doubt the costume is to doubt God. This is what happens in monotheism. If I burn in Hell, it will be for challenging the books of Moses and Abraham. I will plead my case for exercising scrutiny.

"Blessed is the one who does not walk in step with the wicked or stand in the way that sinners take or sit in the company of mockers, but whose delight is in the law of the LORD, and who meditates on his law day and night. That person is like a tree planted by streams of water, which yields its fruit in season and whose leaf does not wither— whatever they do prospers". - Psalm 1:1-3

Prana is real, and LORD technology traps it in its frame. All of our beliefs get sucked into the intake when we confess we're dirty. We subscribe to a gold level service called salvation. But salvation is never a journey forward. Salvation comes on knees. Salvation is a climb out of a pit. The ascension program requires you to be desolate. Salvation promises you will stand where you once stood before you got here. The truth is, most people feel safer in the pit. Shame is a contract with slavery. Sovereignty is the man who truly confesses to himself. Otherwise, salvation is intimidation.

"Serve the LORD with fear and celebrate his rule with trembling." - Psalm 2:11

Inversion splits the mind. It uses the hatred of self to crack open the husk. It empties our pearls for the machine to mince. Abandonment is the origin of inversion. It turns Judaism into blackmail. Catholicism into pedophilia. Christianity into Zionism. Inversion is a psychological disassociation. It exists between the Old Testament and the New Testament. The two towers of mind control are severity and mercy - Jachim and Boaz. These are spark gaps crackling their way up Jacob's ladder. This is the hypnotism of good and evil. The golden child is anointed publicly to make each witness a scapegoat.

When all the people were being baptized, Jesus was baptized too. And as he was praying, heaven was opened and the Holy Spirit descended on him in bodily form like a dove. And a voice came from heaven: "You are my Son, whom I love; with you I am well pleased."
- Luke 3:21

The LORD of the Bible and the LORD of the Quran are the same. LORD is the shepherd, and man must learn to be a good beast. Mankind accepting salvation through the eye of providence is that contract. The New World Order is the blueprint for globalism. This plan will only be forced on a minority. The majority will call it God's will or beg for it to happen.

This is a hypnotic spell older than history. It's a vibration planted over five-thousand-years ago. The only cure is to cut our engagement from its belief. Doing this will feel like a betrayal. This is the adhesive of mind control doing its job. There is a cost for your sovereignty. To be free, one must commit to the quest.

We fear the LORD. The Bible won't even spell out his name. The authors decided our eyes were too profane to read it. Our salvation is placed too deep in the pit to ask for LORD's identification. We give LORD all of our power in exchange for acceptance and mercy. This is the blueprint of mind control. A snake oil runs in the scripture. LORD technology comes with the side-effect of death for the individual.

There can be only one LORD.

The third temple will mask the third world war. A grand reset will subdue and distract us in the Jubilee. All nations will be brought to their knees before LORD. He is the shepherd in the Eye of Providence. In this Apocalypse, the gods of Yahweh and Allah will be revealed as one. All people of the world will be declared children of Abraham. The chosen ones will administer LORD's government like a Tabernacle in the Wilderness. The New World Order is a coming together of the hive. Rapture is our welcome to oblivion.

Globalism is a pyramid scheme. It's a powerful spell of hypnotism. God is centralized behind a chain of command. A central book forms a single eye for seeing. Sixty-six chapters are the entire Alpha and Omega.

Globalism is a trance between pillars. The pillars of the Bible are two testaments. The pillars of Abraham are Islam and Christianity. The pillars of governance are Capitalism and Communism. The pillars of sovereignty are Master and Slave. The pillars of politics are Republican and Democrat. The Two Towers are an ancient spell of sound. Two tones produce a third sympathetically as the resonance splits us open.

Possession is the one true goal of globalism. Possession comes out of disassociation. The two towers are instruments of evacuation. In the Old Testament, God hates you. In the New Testament, he loves you. The dissonance comes in the fluctuation. This is the art of gaslighting.

When one man saves another from his death, we call that a life debt. The debtor's sovereignty is obligated to his master until it can be repaid. Since early Rome, the title, Pater Patriae, was given to the rescuer. This is second father. It's a belief technology made from reverence instead of fear. We earn LORD's mercy by accepting our life debt to his son. In LORD technology, you are a genetically profane being and were saved by the death of the golden child. An oath was sealed in blood on your behalf long before you ever got here. This is how you possess a heart. It becomes too uncomfortable to remain in the chest. We make room for the Holy Spirit.

Stockholm syndrome is trance that can be passed down generationally. Our soul is split with the fork and spoon of the LORD. The fork of guilt pricks us open, the spoon of salvation scoops us hollow. LORD is a technology of belief and possession. The Holy Spirit is an ethereal corporation.

This isn't a call to abandon the Bible. It's a call to reclaim your life. Belief is the most valuable commodity in the world. Its collection has been centralized and used for nefarious purposes. The system is selling you back a tiny portion of your original power after it's used for corruption.

I was terrified to reject the contract of sin. I was alone in the dark pit but found footing with my back against the cold mud wall. But I was tall enough to shimmy up its sunken well. I told God I don't believe in his divine bloodline, and I rose. I told him having a golden child reduces the rest of his children to scapegoats and I climbed.

I dig my heels to shimmy up the dark chimney saying, "Having chosen people is evil." Telling God how you feel is hard. You can feel LORD technology listening. It's as hard as questioning the Holocaust. These struggles are guilt-ridden by design.

Alchemy can only be hard. I clutch the roots of trees as I rise. I know God wrote his law in the stars instead of delegating it to a language. Everyone sees the sky. No one needs a secret language to read it. Stars are a timeless infallible scripture no one misses. If God wanted us to have commandments, he'd be sure everybody sees them. He'd make the moon play a recording in your head when you look up. "Welcome. This is your owner's manual. Touch index fingers to access the table of contents…"

Salvation's pit is not a technology of God. Truth digs its toes into wet mud to find a rock. God knew there would be liars. He invented the tongue. He gave us the tools to be fooled and the tools to notice. God is mining our ability to solve him. There can be no emergency cord. The only way out is to climb.

I spit dirt from my lip and wick wet hair from my eye. I look up through sweaty green eyes. I am jealous of the grass basking. This pit is an opening now that keeps getting bigger. I can see the room at the top. It's the size of outside. Belief is the solo climb. God can't tap on the glass and point to our feet if we drop something.

When I was seven, I ran the soccer ball past my goalie and scored against my own team. No one tried to stop me. Our goalie was dumb-founded. In my mind, I was Pele saving the universe. My mom was our coach. She watched the whole thing patiently. She didn't stop it from happening. If she was yelling, I probably thought she was cheering. That was my first and last goal of the season. I still remember the rush of victory before the truth hit. That's where we are now. We're certain we've won the day, but God won't confirm. This is our game. This is our free will.

The best mind control is a diversity of sound and color from all sides. It hides the fence around the energy ranch we share. The roof of the cage is the space program. The floor is the fear of nuclear weapons. The walls and furniture are health, wealth, information, and government. Our prana is trapped in a cube of many colors. The answer is not to keep reading. Your answer must stand without crutches. Faith is impossible with training wheels. You abandon God's gift by carrying LORD's water. The LORD and God are not the same. That's a side effect of the illusion of one.

Salvation is an agreement to slavery. We have to stop falling for the mind control.

The spell began inside a robe. The first white vestments were worn by priests and doctors. The first black robes were worn by judges and kings. Law is belief invoked by a costume. Law is the spellcraft of a black nobility. Under maritime law, when a ship enters a port, it presents its berth certificate. A child comes to this world through a berth canal. As gematria hides numbers in letters, language hides contracts in sound. A black law claims ownership of the sea and everyone who lives on its islands. It claims us all as children of Rome. Cannibalism is the act of ownership, followed by consumption. Evil extracts consent before sinking its teeth.

The Vatican musters no army. It holds no nuclear missiles. It has no Air Force or Navy. It has no Special Forces. The Vatican is a living triple crown. The world is ruled by a secret triumvirate in the open. Rome rules the heart by conquering the spirit. London rules the mind by conquering the gold. D.C. rules the body by military force. The Vatican claimed ownership of every corpse in every port on November 18th, 1302. Pope Boniface VIII issued the decree Unum Sanctam. Outside of the Church, there is no salvation.

"We are obliged to believe and confess with simplicity that outside the Church there is neither salvation nor the remission of sins." - Pope Boniface VIII

The Pope would cleanse what he owned. Boniface was casting the spell, plenitudo potestatis (plenitude of power). Those who resisted the Roman Pontiff were resisting God's ordination. Boniface explained, "in this Church and in its power are two swords; namely, the spiritual and the temporal." Since the body is governed by the soul and the soul is governed by the spirit, the Roman Pontiff is governor of both soul and body.

"We declare, we proclaim, we define that it is absolutely necessary

for salvation that every human creature be subject to the Roman Pontiff." - Unum Sanctam.

In 1455, the decree Romanus Pontifex established the divine right to all property:

"The Roman pontiff, successor of the key-bearer of the heavenly kingdom and vicar of Jesus Christ, contemplating with a father's mind all the several [climates] of the world and the characteristics of all the nations dwelling in them and seeking and desiring the salvation of all, wholesomely ordains and disposes upon careful deliberation those things which he sees will be agreeable to the Divine Majesty and by which he may bring the sheep entrusted to him by God into the single divine fold, and may acquire for them the reward of eternal felicity, and obtain pardon for their soul."

After dehumanizing a target, there is no need to give it rights. The Holy Church, like the narcissist, is entitled to special privileges:

"to invade, search out, capture, vanquish, and subdue all Saracens and pagans whatsoever, and other enemies of Christ wheresoever placed, and the kingdoms, dukedoms, principalities, dominions, possessions, and all movable and immovable goods whatsoever held and possessed by them and to reduce their persons to perpetual slavery."

Rome would later be the legal basis of Christopher Columbus's authority as a viceregal agent of the Spanish crown to claim the Americas. Rome was the birthplace of the legal system and ecclesiastical spell-craft. Court is the first ritual circle. All words are bound in stone under penalty of death. We swear on a black Bible before we take the pulpit in the black court. Like any religion, the courts are a belief technology. The hearings are the confession. The sentence is penance. The jury is the apostles. God's judgment, channeled through a man in a black robe, is singular and immovable.

Rome did not fall; it went underground and transformed. Constantine was fighting paganism. Its roots were decentralized. This made it impossible to control. He had to break belief to capture their spirit. A trident is needed to rule the world. Capture the body, mind, and spirit simultaneously. This is the purpose of the Vatican's outer church. This is the church of citizenship. It includes all of Christianity. We never see the inner church. Like the Holy Tabernacle, only the high priests are allowed inside. They wrote a religion specifically for the outer church to serve the Tabernacle's purpose. The children of Rome are the Tabernacle's untouchables.

Almost all of the Biblical history of Jesus comes from Josephus. Josephus Flavius served under Titus Flavius, the Roman emperor who "raised the temple" at Jerusalem in 70 AD. Titus was called a messiah in his time. According to writer Joseph Atwill, the gospels foretold the son of man would do three things: One. Crush a Galilea Town. Two. Encircle Jerusalem with a wall. Three. Raise a Temple. Titus Flavius did all of these things forty years before the gospels.

Belief is the magic of Egypt. Its name means the land of black. A Black Nobility still sits atop the pyramid. The city of Heliopolis is mentioned in the Bible but never the Great Pyramid. These two places are less than twenty miles apart. Belief is a commodity subject to the powers of monopoly and deception.

"When Israel was a child, I loved him, and out of Egypt I called my son." - Hosea 11:1

A centralized belief technology claims a single divine Pope. It promotes humans into Saints for good behavior. The Roman Catholic Church built its sanctuary around the human remains of St. Peter. The Basilica's foundation rests on a covered cemetery. The Pope is the arbiter of miracles. The church embeds human bones in every pulpit of every new church they open. This is the technology of demons. They are conscripted to service the LORD or return to the pit. LORD technology tells you to eat the body and drink the blood in remembrance. A living altar is set in your heart as you contemplate the suffering. There is prana that flows from your knees in front of a crucifix as you feel sorry. It empties your pockets spiritually. You surrender to the weight of guilt that wasn't yours.

Who owns the Vatican trust? Who gave us the Bible? Who told you God sacrificed his son in a torturous ritual of crucifixion to absolve you of evil? What does this false-guilt do to your psyche? Does it make you feel bad? Would you do anything for absolution?

The Blueprint for Mind Control:
   Step 1.) Target is a sinner.
   Step 2.) Target needs external help.
   Step 3.) Salvation is traded for devotion.

The Gospels of Mary and Thomas speak of self-empowerment. The church censored both books. Christ is a channel you tune with your body. When your spine tunes the Cross inside, your body resonates a divine signal. You glow from the place of milk and honey. The third ventricle is found between the pituitary and pineal. The Two Towers produce a sympathetic third tone between them.

If you vibrate guilt or shame, you introduce dissonance to the channel. This is the purpose of mind control. The Obelisks of Rome were entrained with a guilty pulse. They broadcast dissonance to block you from the signal.

Let go of their baggage. God is not jealous. God is not vengeful. God doesn't need our attention. God does not require our obedience. God doesn't insert himself into our culture through a committee of pedophiles in Rome. God doesn't even mind if we want to be ruled by a psychopathic cult of liars. God gave us sovereignty. God respects our adventure. Attaching a blood oath or a life debt goes against the architecture of this world. We're not born with hooks in our back. You don't need a golden key of privilege to return to his building. Would you do that to your children?

God lacks the capacity of jealousy. That's why he made us. We are the synaptic feelers here to document the journey from lack. God is a responsible parent. He respects the rules of the aquarium. He knows he disrupts the water by sticking his hand inside. God doesn't need to change our course because that would imply we surprised him. Surprise is an illusion of time.

The Flavians created a sin cult. They needed you to feel unclean the moment you were born. They need you to believe you are unsanitary, meek, and guilt-ridden. They need you to believe in death and taxes. By 1905, they no longer needed you to believe in sin. Instead, they convinced you there was no God. Atheists believe in death, science, and taxes. This makes for a perfect sharecropper to shed his labor willingly. Stepping out of the sin cult is brutal. The shame is a perk of LORD technology. It's a self-inflicted burden requiring all of your prana.

Black magicians use the power of leverage and implication in sinister ways. The son of God is exclusionary. It makes everyone's bloodline unholy. If you worship Jesus Christ you are not worshiping God. You are worshipping the son. The word Christ means anointed. Jesus was the anointed one. Jesus was the chosen one. Jesus is the archetype of the golden child.

Rome isn't the Pope. Rome is the Vatican trust. They own the Queen's monarchy, and America is an extension of her Royal Navy. The Royal crown is the Holy Crown. The original meaning of the trinity. Father, Son, and Holy Ghost. Mind, body, and spirit. Financial, Military, and Religious.

The Jesuits founded in 1541. The globe appeared in 1543. Jesuits invented globalism. Pietro Pomponazzi and Gasparo Contarini took greek philosophy and made it into a corporation. Globalism was the birthplace of religion. Before that, belief was derived from Philosophy. Religion is victory. Philosophy is wisdom. LORD technology insists there can be only one. All of the following theories are Jesuit inventions of mind control: Copernican model, Newtonian Gravity, Darwinian evolution, Einstein relativity, and Big Bang Theory.

The Council of Trent was an 18-year corporate board meeting. Dress codes, music, scripture, and even the economy of the church job market were fed into the bureaucracy of worship. The church's inquisition was the acquisition of spiritual obedience. Bureaucracy is the energy of chain-of-command. It grows in the evacuation of autonomy. One has subjugated its sovereignty to a higher power. That power is one of three obelisks in a global triumvirate. A corporation selling salvation grows by tweaking supply and demand. In the devil, the church will be a prime solution. Add witch burnings, and the church invents the first false flag.

Mythraism was the religion of Rome. Mythra, the Sun, died on a cross with twelve followers from a virgin mother. Peter and Constantine transmuted Mythra in Christianity. Christianity goes back to the zodiac and the city of Petra. Peter hijacked the rock for the LORD. LORD technology is not God.

We are living Christians. It's so much more personal than sun worship. As walking antenna, we tune God. The Bible has distortion. Remove the wizardry of bloodline. Remove the reverence for torture. Remove the Passover sacrifice. Remove the obedience to government. Remove the idolatry of relics. Remove the politics of Constantine, Rome, and Peter. What's left of the Bible is Jesus, the Good Samaritan.

The Good Samaritan was a real man. He was a walking Christ. He found a way to tap into his sacred fluid. The Samaritan was beaming with joy and wandered the lands embracing strangers. He was walking in the Way of Christ. The Cross. The "Sun" of God is Mythra. The slayer of the bull and the end of the age of Taurus. Ancient belief was tracking the stars. They tuned into the scripture of creation. A sacred bloodline is the racism of inbred Jesuit psychopaths. If Jesus is the only son of God, how can we be God's children? A sinister pyramid was inserted between you and God. The LORD is belief technology.

In 1000 BC, Mithras was born on Dec 25th to a virgin named, Anahita. Anahata is the heart chakra. This is the sacred heart. Mithras said each man was a divine "son" of the One God.

"Do unto others as you would have them do unto you." - Mithras, 1000 BC

This was early Christianity. It came with no bloodline mythology. The original way was the Zodiac. God's true Bible was and has always been the stars. Something we don't see that much of anymore. We gaze into the false light of black mirrors while God chose the sky as his scripture. No Pope could ever corrupt the sky. They can only corrupt what we think it means.

The quest for truth is the practice of alchemy. A fruitful search is a transmutation of lead into gold. To decode the truth is to unveil the propaganda in silk. The work of dark to light is a revelation pretending to be a dark apocalypse. Naked goosebumps hide behind the horsemen of Death, Famine, War, and Conquest. Black priests have reinvented the ancient story of Mithra. In the first century, the four horsemen were the cardinal directions of the sun. They looked. And a White Horse rose in the east as dawn conquered the dark. Behold, the Pale Horse rides in the clear light of truth. As a Red Horse went out, the setting sun took peace from the earth. For in the sun's death, the Black Horse rendered blind justice. Take heed to not damage the oil and the wine, for the White Horse will rise again.

This story is the chariot of Mythra. The cycle of the sun was the worship of light's resurrection. The light dies to be reborn. This is astrotheology. The original Christianity now locked behind seven seals. We have to unwrap the Bible like a mummy. Truth is a mighty Sequoia wrapped in a choking vine of thorns. That vine is ancient black magic. Revisit the archetypes of Kane and Abel. Were they good and evil? Or sunrise and sunset?

Astrotheology is the language of day and night. Roman cults are the language of good and evil. Before Rome, messiahs were prolific and dangerous. Messiahs were trendsetters with a lot of charisma. Constantine centralized religion in Nicea. He took every messiah, past or future, and combined them into One LORD.

Why would a Roman dictator declare there was only one messiah? Imagine thousands of radio stations across the country, replaced with a single channel. Who would want that more - God or Satan? Reject the First Triumvirate of Caesar, Crassus, and Pompey. Reject the Second Triumvirate of Antony, Octavius, and Lepidus. Reject the Third Triumvirate of Rome, London, and D.C.

Let this neither dissolve or threaten your relationship with God or Jesus. We must remove the noise separating us from God. That is our duty as living crosses. Dismiss all that binds you to a blood oath or a life debt. Jesus is the Good Samaritan - the first truther.

Thomas Sheridan reminds us of the first Cross. God's communion with man was the gift of fire. God made us all Prometheus in the rubbing of sticks in the perpendicular. God punishes no one for lighting his plasma. On the contrary, it's why we're here. The Cross is a symbol of inner light long before it was a symbol of bloodline or crucifixion. It's a truer meaning of man's savior than any suffocation ceremony on a shrine.

LORD technology has grown since Rome. It jumped from Rome to London via the Venetian Empire. A new kind of belief was born. The war for the soul fell into the clutches of materialism. New scriptures were crafted by Isaac Newton, Voltaire, Pomponazzi, and Russell. The plot thickened as the budget for narratives grew.

Equality was a psyop of the French Revolution. The first yellow vest protest in France was July 14, 1789. Washington, D.C. was founded on July 16, 1790. London and D.C. were installed as power centers the same year France fell. The history we are told is mythology. The New World Order was taking over.

Science is the new religion. The shame of sin is replaced with the shame of carbon footprints. We are as dirty as we ever were but for different reasons. Which scenario is more likely? We dumped all of the Queen's tea and defeated the most powerful military power in the world, or a mega government was formed?

Jesus was co-opted by LORD technology. LORD made Jesus the crook and Satan the flail. Herod killed every firstborn male under two years old to taint family inheritance with trauma. This is called survivor syndrome. The victim rejects his own life and fortune from the trauma. The goal was to create an overarching sense of authority. The technology of walls comes with the belief there are no wolves inside.

The coat of many colors is a kaleidoscope of shadow. LORD is the single shining pyramidion in the desert. Excommunication from the church is the narcissist discard. The Divine Eye of Providence is the science of love-bombing. Hell is the scapegoat. Heaven is the golden child. These are both self-abandonment. The true ailment we see today in our belief. We have abandoned our personal technology. We have installed LORD technology in its place.

Like the Quran, the Bible has been tuned for a purpose. God invented sovereignty and natural law. The technology of LORD invented central authority. Nature has no central authority. Nature is naturally decentralized. Much of the Bible claims the opposite:

"The authorities that exist have been established by God. Consequently, whoever rebels against the authority is rebelling against what God has instituted ... Therefore, it is necessary to submit to the authorities, not only because of possible punishment but also because of conscience. This is also why you pay taxes, for the authorities are God's servants, who give their full time to governing." - Romans 13:1-7

Jesus was a revolutionary. Yet in Mathew 22:21, he condones a city built on slavery, "Render unto Caesar the things that are Caesar's, and unto God the things that are God's." This isn't the only place in the Bible where we are told government is a divine authority.

"Submit yourselves for the Lord's sake to every human authority: whether to the emperor, as the supreme authority, or to governors, who are sent by him" - *1 Peter 2:13-14*

Both the Bible and LORD technology condone slavery. The same is true for the Quran and Islam.

"fear God, honor the emperor. Slaves, in reverent fear of God submit yourselves to your masters, ... To this you were called, because Christ suffered for you." - *1 Peter 2:17-20*

Either the whole Bible is true, or it is corruptible. Satan is credited with poisoning everything in this world except the scriptures. He had plenty of chances but was unsuccessful through the scripture's five languages, dozens of Popes, emperors, royals, and the bureaucracy of the Nicene Creed. It is delusional to think Satan exists but failed to penetrate the Bible. Nowhere is Satan more successful than inside a coercive state or religion. Both are centralized pyramids of authority.

The Bible suffers the same corruption as any technology. Government is a false idol. Pull the hood off the Apocalypse. The pyramid will only grow taller if we don't. False gods have been placed on the mantle as we toil to please them. It's time for necromancy. You serve God through your devotion and reverence for sovereignty. Nothing you could do would be more godly.

CHAPTER THIRTY-THREE
Apocalypse Now

A leviathan cut its teeth in Egypt. It struggled its way north through Rome and into Venice. It invented the royalty of blood and the spiritual monarchy. It charged the word "Saint" with sanctity. It infiltrated the Church, the Crown, and every corporation. It grows fat from our denial.

In WWI it formed the League of Nations. In WWII, it formed the United Nations. This creature grows in our cowardice. We demand a new world order as the old one burns. Every war cements belief in a shiny new temple. Every dictator sells the necessity for oppression. Every day we complain these chains aren't tight enough. We place ourselves below the elements of silver and gold. We believe every lie we've been told. The children of Rome are masters of their own slavery.

A good news is rising. Apocalypse is an uncovering. Most don't stomach what's underneath the sheet. We blame the devil instead of psychopathy. The truth is so simple when you're no longer in fear. Five thousand years ago, kabbalists wrote this script. We must not mistake the New World Order for God. If we do, we'll end up owing him an apology.

We keep swallowing lies. In 2001, the US invaded Afghanistan, and two years later, they established a central bank. In 2003, the US invaded Iraq, and the next year they established a central bank. In 2008, Venezuela paid off all its loans to the World Bank. Three years later, their President accused the United States of germ warfare. In 2017, Venezuela stopped using the petrodollar. A year-

and-a-half later, America called their President illegitimate. In 2009, Hillary Clinton did the same thing to Honduras. In 2011, the United States bombed Libya and established a new Central Bank while Kadafi was still in power. The architect of the world's largest freshwater irrigation project was assassinated for not bowing to the beast. America has always been the NWO. America has always called what it does liberation.

Uncovering the truth is painful. We are subdued by the denial of who we are. We are outcasts from the garden of morality. Banishment gives us the freedom to destroy this world. We cherish the gears that shake the beacon of democracy. We watch them crush the bones of liberty and self-preservation. General Wesley Clark told us which countries we'd attack in what order. Still, we trudge on. We are henchmen to the Eye of Providence. We are the muscle of the New World Order. The Apocalypse could not come faster.

The oldest secret society we know rose from Kabbalah. The first man, Adam, was the first magician. Adam, Jacob, and Isaac were all kabbalists. It was an ancient mesmerism that cast us out of the garden. The first lie cast by the first magician. Two angels bearing flaming swords perched outside the nightclub of Heaven. The angels believed, "God told me to tell you no one gets in." Our salvation is commandeered by bouncers and Peter was the first gatekeeper. God banished no one. The infinite lack the notion of exclusion.

Before the first lie, deception was a virgin. The sanctity of lips was so pure it curled the feathers of peacocks. The telling of a lie is the first black magic. The tails of snakes rattle fear into obedience. We banish ourselves when with our lies. We become our own secret society. We think everyone speaks the truth, so people will be more prone to believe. We know not what we do for strategic reasons.

An ancient meaning whispers behind a veil of forked tongues. Kabbalah is a vocal tradition pretending to know God's name. First, they called him Zoroaster. Then they called him Uhura Mazda. Then they called him Yahweh and Allah. Only you know God's name. Don't give that power away.

God's song is the unanimous sound of every voice in silence.

Special thanks to my Patrons who made this book possible: Aaron Beattie, Adam Duncan, Adriana, After Burner, Arlene Graham, AYK, Beth Martens, Bill Craig, Brent Scheneman, Brett Denbow, Casey Congdon, Callista Summerfield, Christine Vincent, Colleen Casey, Dan Malcolm, Diane Polzer, Don Peterson, Harvey Browne, Heidi, Henk de Vries, Isla Swan, Jarod Andrew Johnson, Jeff Gates, Jennifer Mazgelis, Jill Wong, Joe Ingenito, JK, Joseph Sullivan, Joshua Smith, Juggy, Jugoslav Vukicevic, Juliena Sharp, Kimberley Fisher, Leslie Jones, Lori Wince, Mary Krombel, Michael Jaeger, Neilly, Pam Hunsicker, Paul Farace, Peter Westberg, Phil Taylor, Pinkema Artiste-Model, PurplePetunias, QC, Raymond Smith, Rebecca Antonelli, Sebastian Ostrowski, Shannon DiGirolamo Bates, Steven Mercer, and Zay.

Many aren't mentioned. To everyone who's read my book, column, listened, watched, recommended, commented, shared, liked, subscribed, invited to speak, reached out, or sent letters I sincerely thank you for the support. I wish I could mention you all by name. I have much gratitude.

Printed in Poland
by Amazon Fulfillment
Poland Sp. z o.o., Wrocław